Supercritical Fluids Technology in Lipase Catalyzed Processes

Supercritical Fluids Technology in Lipase Catalyzed Processes

Sulaiman Al-Zuhair
and Hanifa Taher

CRC Press
Taylor & Francis Group
Boca Raton London New York

CRC Press is an imprint of the
Taylor & Francis Group, an **informa** business

CRC Press
Taylor & Francis Group
6000 Broken Sound Parkway NW, Suite 300
Boca Raton, FL 33487-2742

First issued in paperback 2017

© 2016 by Taylor & Francis Group, LLC
CRC Press is an imprint of Taylor & Francis Group, an Informa business

No claim to original U.S. Government works

ISBN-13: 978-1-4987-4387-7 (hbk)
ISBN-13: 978-1-138-89320-7 (pbk)

Visit the Taylor & Francis Web site at
http://www.taylorandfrancis.com

and the CRC Press Web site at
http://www.crcpress.com

To my mother, wife, and my beloved children, Abdulrzak and Ranya. Thank you for your support and love.

I also dedicate this work to my late father—I wish you could read the book and give me your valuable feedback. I miss you so much.

Sulaiman Al-Zuhair

Contents

Preface

Enzymes are catalysts of biological sources, and, like other catalysts, they bring reactions to equilibrium more quickly than would occur in their absence. They are proteins, and, therefore, understanding enzymes requires the understanding of proteins structure and their environmental behavior. Enzymes are used on a daily basis and can be found in the food, pharmaceutical, detergent, and many other industries. The book gives basic information about enzymes, their sources, reaction kinetics, and main industrial applications. Techniques of isolating, extracting, and purifying enzymes are also presented.

The book looks deeper into lipases, which are hydrolytic enzymes that have great potential in many industrial applications. The main sources, structure, and features of lipases are presented, with an emphasis on their specificity and interfacial activity. Various industrial applications and property improvements are also discussed. The book focuses on lipase immobilization and its advantages over soluble lipases, where its effects on the physiochemical characteristics, namely, activity and stability, of lipases are discussed. Different immobilization techniques are described, and examples of various immobilization materials are given. In addition, different bioreactor configurations using immobilized lipases are described.

The book also discusses the advantages of nonaqueous media in biochemical synthesis over aqueous and solvent-free systems. The focus of the book is more on the use of supercritical fluids (SCFs) as a green alternative reaction medium. Factors affecting the physical properties of lipases in this medium, and hence their activity and stability, are discussed. A case study using supercritical carbon dioxide (SC-CO_2) for biodiesel production is presented. Biodiesel, derived from vegetable oils or animal fats, represents a promising alternative fuel for use in compression–ignition (diesel) engines. It comes from renewable sources, is biodegradable, and less toxic, as it is not petroleum-derived. Compared to petroleum-based diesel, biodiesel has a more favorable combustion–emission profile, such as low emissions of carbon monoxide, particulate matter, and unburned hydrocarbons. Lipases have been suggested as a better alternative to conventional chemical catalysts for biodiesel production. At present, the high cost of biodiesel is the major obstacle to its commercialization. The book presents the new, cutting edge technology, using enzymes to reduce the overall production cost.

In summary, the book provides the reader with information on the available technologies in lipolytic reactions. The book also provides useful information on the use of lipases in nonaqueous media to overcome the drawbacks usually encountered with the use of conventional chemical catalysts. Immobilization techniques and the use of immobilized lipases that allows repeated use, which is essential from economic point of view, are also covered in this book. The book gives the reader a clear idea on the challenges and new frontiers in the field, including the use of SC-CO_2.

Acknowledgments

We thank Dr. Ali Al-Marzouqi, Professor Yousef Hayek, and Professor Mohammed Farid for their valuable contributions to this work.

Authors

Sulaiman Al-Zuhair is a professor of chemical engineering at United Arab Emirates (UAE) University. Prior to joining UAE University in 2006, Al-Zuhair worked as an assistant professor in the School of Chemical and Environmental Engineering at Nottingham University, Malaysia Campus. He obtained his PhD in biochemical engineering from the University of Malaya in 2003, his MSc degree in chemical and environmental engineering from University Putra Malaysia in 1998, and his BSc degree in chemical engineering from the Jordan University of Science and Technology in 1996. Al-Zuhair has published more than 50 peer-reviewed journal papers, 2 patents, and 2 book chapters. He is a member of the editorial boards of several international journals. The majority of his research work is on the uses of enzymes in various industrial applications, specifically in biofuel production.

Hanifa Taher is an assistant professor of chemical and environmental engineering at the Masdar Institute of Science and Technology. She earned her PhD in Chemical Engineering from the United Arab Emirates (UAE) University in June 2014. Taher's research work is focused on the enzymatic production of biodiesel from lipids extracted from different feedstocks using supercritical carbon dioxide (SC-CO_2). Taher developed a novel integrated continuous biodiesel production process in SC-CO_2. Taher has reviewed a number of research articles submitted for publication to international journals including *Biomass and Bioenergy*, *Fuel*, *Journal of Supercritical Fluids*, and *Waste Management*.

1 Enzymes Fundamentals

Enzymes are catalysts of biological sources. Like other catalysts, they bring reactions to equilibrium more quickly than would occur in their absence. They are proteins and, therefore, understanding enzymes requires an understanding of protein structure and their environmental behavior. This chapter covers basic information about enzymes, and discusses their sources and main features before introducing their classifications based on International Union of Biochemistry and Molecular Biology (IUBMB) standards. Techniques for isolating, extracting, and purifying enzymes are also presented. The chapter concludes with the main industrial applications of common enzymes and genetic protein engineering.

1.1 CATALYSTS AND ENZYMES

Chemical reactions usually require energy, in the form of heat, to make particles move rapidly and increase the collusion frequency between them. The energy required to convert one mole of reactant molecules from their stable state to a transition state is known as the activation energy. As molecules collide with a proper orientation, some bonds are broken and others are formed. The required energy can be lowered by adding catalysts, and, thus, reaction rates are enhanced.

Catalysts are substances that alter the rate of a reaction without being consumed during the reaction. Generally, catalysts increase the rate of reactions by providing alternative and faster paths through which the reaction can proceed. In every chemical reaction, reactants absorb energy and pass through a transitional state between the reactants and the products, thus forming intermediate compounds. These intermediate compounds immediately decompose to give the end product. The catalyst's role is to reduce the energy required to reach a transitional state by bringing the reactants closer and to allow them to collide more effectively. However, catalysts do not affect the reaction equilibrium. This means that they accelerate both forward and backward reactions by the same factor. Figure 1.1 shows typical reaction progress in the presence and absence of a catalyst. As can be seen, using a catalyst reduces the energy barrier between the reactants and the transition state.

Although chemical catalysts are important and have many industrial applications, their use usually has negative environmental and economic impacts. In addition, they usually result in an increased number of by-products, which must be separated from the desired end product.

Alternatively, enzymes, which are biocatalysts, have many favorable properties compared to conventional chemical catalysts. Enzymes are large proteins that are made up of a sequence of 20 naturally occurring amino acid residues to form a chain connected by peptide bonds, shown in Figure 1.2. As with chemical catalysts, enzymes alter the reaction by providing different pathways and operate effectively in small amounts (Lorenz and Eck, 2005). It has been reported that enzymes

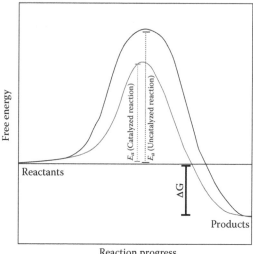

FIGURE 1.1   Effect of catalysts on activation energy.

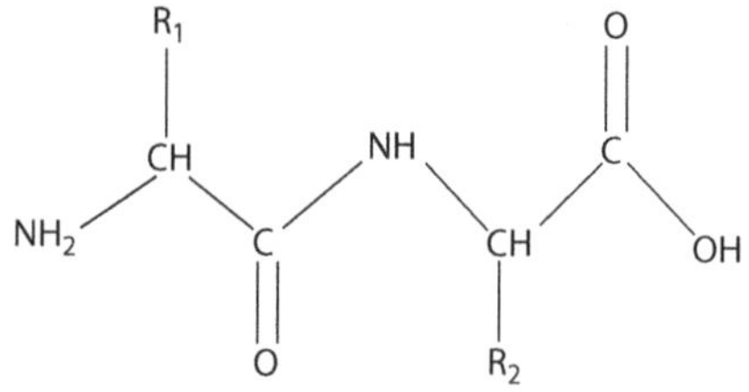

FIGURE 1.2   Peptide bond in a protein.

are capable of enhancing reaction rates by up to 10^8- to 10^{10}-fold (Freifelder and Malacinski, 1993). Table 1.1 shows the enzyme's effect in reducing the activation energy required for different reactions. The table clearly shows that enzymes have a stronger effect than chemical catalysts.

In addition, enzymes are much more specific and can produce only the desired end products without any side effects. They also work well in moderate pH (pH 5 to 8) and temperatures (20°C to 40°C), which makes them less hazardous and less energy intensive.

Typically, enzyme active sites consist of 3 to 12 amino acid residues structured into specific three-dimensional arrangements, as shown in Figure 1.3 for the *Candida rugosa* enzyme. These active sites have a strong attraction to substrates, where amino acid residues complement certain groups on the substrates. Regiospecificity and stereospecificity are other important characteristics of enzymes, where they are not only specific or selective for certain substrates but also discriminate between similar parts of molecules or optical isomers. For examples, some enzymes, like lipase, can differentiate between stereoisomers that have the same structure but a

TABLE 1.1

Effect of Enzyme Use on Reaction Activation Energy

Reaction	Catalyst	Activation Energy (cal/mol)	Reference
Hydrogen peroxide decomposition	Without catalyst	18,000	Campbell and Farrell, 2011;
	Platinum surface	11,700	Spencer et al., 2010
	Potassium iodide	13,500	
	Catalase	5500	
Ethyl butyrate hydrolysis	Hydrochloric acid	16,800	Steward and Bidwell, 1991
	Lipase	4500	
Sucrose hydrolysis	Hydrochloric acid	26,000	Goss, 1973
	Invertase	13,000	
Casein hydrolysis	Hydrochloric acid	20,600	Braverman and Berk, 1976
	Lipase	12,000	

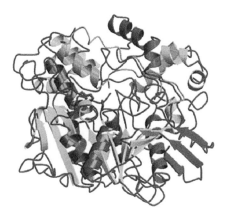

FIGURE 1.3 Three-dimensional structure of *Candida rugosa* enzyme. (From Cygler, M., and J. D. Schrag, 1999, *Biochimica et Biophysica Acta (BBA)—Molecular and Cell Biology of Lipids* 1441 (2–3):205–214. With permission.)

different atom arrangement. Details of lipase stereo- and regiospecificity are discussed in Chapter 2.

Enzymes, on the other hand, have several disadvantages such as high cost of production, isolation, and purification. It has been reported that in enzyme production, 45% of the costs are associated with enzyme recovery, whereas only 14% are attributed to the fermentation process (Fish and Lilly, 1984). Enzymes production and purification are discussed in Section 1.4.

1.2 SOURCES OF ENZYMES

Enzymes can be obtained from different sources such as in the pancreas and liver, plants, and microbial organisms (such as bacteria, fungi, and yeast). Historically,

TABLE 1.2
Common Microbial Enzymes and Their Industrial Uses

Enzyme	EC Number	Origin	Source	Optimum Conditions Temperature (°C)	pH	Industry	Reference
α-Amylase	3.2.1.1	Fungal	*Aspergillus ozyzae*	50	4–5	Food and detergent	Chang et al., 1995; de Souza and de Oliveira e Magalhães, 2010
			Penicillium chrysogenum	30	5	Food and starch	Balkan and Ertan, 2005; Gouda and Elbahloul, 2008
		Bacterial	*Bacillus licheniformis*	38	7	Bacteria, food fermentation, textile, paper, and starch	Ul-Haq et al., 2007
			Bacillus circulans	60	6		Takasaki, 1983
			Bacillus amyloliquefaciens	45	6		Lee et al., 2006
β-Amylase	3.2.1.2	Bacterial	*Bacillus cereus*	37	6	Starch	Hirata et al., 2004
			Bacillus megaterium	60	6.9	Food	Ray, 2000
Glucose oxidase	1.1.3.4	Fungal	*Aspergillus niger*	40	5.9	Food	Constantinides et al., 1973; Kona et al., 2001
				40	5.5	Pharmaceutical	Szajáni et al., 1987
			Penicillium amagasakiense	40	4.5	Food preservation and chemical synthesis	Kalisz et al., 1997
Hexose oxidase	1.1.3.5	Algae	*Chondrus crispus*	–	–	Paper and food	Bartlett and Whitaker, 1987
Cellulase	3.2.1.4	Bacterial	*Thermobifida fusca*	50	–	Paper, food, textile, detergent, and pharmaceutical	Deng and Fong, 2010
Lactase	3.2.1.108	Bacterial	*Rattus norvegicus*	25	6	Food	Hermida et al., 2007
Lipase	3.1.1.3	Bacterial	*Pseudomonas aeruginosa*	40	8.9	Food	Finkelstein et al., 1970
		Fungal	*Candida antarctica*	50	–	Biofuel	Pilarek and Szewczyk, 2007

enzymes were initially obtained from animals such as pigs and cows, then from plants, and finally from microorganisms (Panesar et al., 2010; Shanmugam, 2009).

Microbial enzymes are extracted from fermented bacteria or fungal organisms. Recently, microbial enzymes have been used with increasing frequency. This is because microorganisms can be easily and quickly grown on a large scale. They are relatively inexpensive and provide a continuous and reliable supply as compared to animal and plant sources. Microbial enzymes represent about 90% of all commercially produced enzymes (Godfrey and West, 1996; Margesin et al., 2008). The best known sources of microbial enzymes are *Rhizopus nerveus*, *Bacillus licheniformis*, *Aspergillus oryzae*, and *Aspergillus niger*. Table 1.2 shows examples of important microbial enzymes, their sources, and industrial uses. Further details on these industrial uses are discussed in Section 1.5.

1.3 CLASSIFICATION OF ENZYMES

An enzyme is often named by adding the suffix -*ase* to the name of the substrate it works on in nature. For example, the first enzyme discovered was urease, which catalyzes the hydrolysis of urea producing ammonia and carbon dioxide. However, some enzymes have been given uninformative names such as catalase, which catalyzes hydrogen peroxide to water and oxygen. The IUBMB has adopted standards for enzyme nomenclature according to the type of reactions they catalyze and the substrates acted upon.

Six types of reactions catalyzed by enzymes have been identified, and accordingly the enzymes have been classified into six main groups. These are oxidoreductases, transferases, hydrolases, lyases, isomerases, and ligases. These six groups and their respective main functions are summarized in Table 1.3. Of the six groups in

TABLE 1.3
Six Major Classes of Enzymes

Class	Function	Examples
Oxidoreductases	Catalyze oxidoreduction reactions by adding/removing hydrogen bonds	Glucose oxidase, lactate, alcohol dehydrogenase, laccase
Transferases	Transfer of amino, fatty acid, methyl or phosphate functional groups from one molecule to another	Starch phosphorylase, amylosucrase, dextransucrase, levansucrase, aspartate aminotransferase
Hydrolases	Catalyze the hydrolysis of carbohydrates, lipids, proteins, or phosphoric acids esters by breaking single bond and add water across bond	Feruloyl esterases, lipase, chlorophyllase, α-amylases, β-amylases, chymosin
Lyases	Catalyze the breaking/forming of chemical bonds by means other than hydrolysis	Alliinases, cystine lyases, histidine ammonia-lyase
Isomerases	Catalyze the structural rearrangement of isomers	Xylose, mutase
Ligases	Catalyze reactions by binding two chemical groups to form one molecule with the need of ATP energy	Ubiquitin-protein ligase, D-alanine-(R)-lactate ligase, butyrate-CoA ligase, glutathione synthase

the Table 1.3, only the hydrolases have a significant industrial function (Galante and Formantici, 2003), because enzymes of this type can be easily extracted and isolated. They are also capable of working without the need for a coenzyme. Therefore, they can be produced at a relatively low cost and can operate under harsh conditions. Amongst the hydrolysis enzymes that have received a high level of attention are lipases, due to their structural features, favorable properties, and the versatility of the reactions that they catalyze. Details about lipases and respective features are discussed in Chapter 2.

1.4   ENZYME PRODUCTION

Industrial enzyme production should be practical and economically feasible. The selection of a particular enzyme production process depends on the production level, source of the enzyme, and type of application. For example, enzymes used for medical purposes need to be of high purity (Costa et al., 2005), whereas those used in food and fuel applications do not need such high purity levels. However, they require larger quantities.

Enzymes can be produced by submerged fermentation and recovered either from the fermentation medium, extracellular enzymes, or from the cell or intracellular enzyme (Geciova et al., 2002). Generally, the microbial enzyme production process is divided into (1) enzyme synthesis, where cells are produced; (2) enzyme recovery, where the enzymes are extracted from the cells produced; and (3) enzyme purification, where contaminants are removed. These steps may then be followed by enzyme formulation to reach the final desired form.

The majority of the enzymes produced are usually retained within the cells, and only a small proportion may be released into the surrounding environment. The retained enzymes need to be isolated, released, and then purified. To simplify the isolation and purification steps, commercial enzymes are produced from cells that generate extracellular enzymes. The selection of enzyme-producing cells can help in this regard. Extracellular enzymes from fugal (*Aspergillus* sp.) or bacterial (*Bacelluis* sp.) sources are the leading commercial enzymes. Nevertheless, commercial enzymes that are intracellular are also found, such as glucose oxidase and catalase from *A. niger* and invertase from *Saccharomyces cerevisiae* (Chisti and Moo-Young, 1986).

As mentioned earlier, microbial enzymes are produced by submerged fermentation, in which a bioreactor is filled with medium and inoculated with specific microbial cells. The mixture remains under controlled conditions until a sufficient level of production is obtained. If the enzyme is intercellular, the cells need to be harvested for enzyme recovery; but in extracellular enzymes, this is not required. Some enzymes can be intracellular in one organism and extracellular in another. For example β-galactosidase (lactase) is extracellular in *Penicillium* sp. but intracellular in *Kluyveromyces marxianus* (Stred'anský et al., 1993). In addition, intracellular enzymes can be changed to extracellular by genetic and protein engineering modifications (Geciova et al., 2002).

Once the enzyme is generated, the cells need to be separated. Cell separation can be performed by centrifugation, especially for small cells that require flocculation. However, centrifugation has some drawbacks such as the generation of heat that can denature the enzyme. If the enzyme is extracellular, then enzyme recovery is from the supernatant; whereas it is from a solid phase in intracellular cases. For intracellular enzymes, separation can be carried out effectively using depth or tangential flow filtration, which allows for washing the cells to increase the yield (Schmauder and Schweizer, 1997).

The foremost difference between extracellular and intracellular enzymes purification is the need for cell disruption. If the enzyme is intracellular, extraction takes place after cell lysis, which adds an additional step and costs to the purification process. This is not required for extracellular enzymes. Once the cells have been broken down and the intercellular enzymes are released, purification techniques similar to those used with extracellular enzymes are applicable. Intracellular enzymes from animal sources can be easily extracted due to the absence of a cell wall, whereas enzymes from plant and microbial cells require more force due to their rigid cell wall structure. Another approach to facilitate intracellular cell isolation is by genetic engineering. Details of genetic engineering are given in Section 1.6.

A variety of cell disruption methods are available. Physical, chemical, and enzymatic methods have all been used. The proper method should be carefully selected to ensure maximum cell disruption with minimum enzyme damage. This depends on the enzyme source, nature and stability.

1.4.1 MECHANICAL CELL-DISRUPTION METHODS

Liquid shearing, such as high pressure homogenization, solid shear, such as bead milling and the extrusion of frozen cells are the most common mechanical methods for cell disruption (D'souza and Killedar, 2008; Hatti-Kaul and Mattiasson, 2003). These methods are mainly based on the use of mechanical force to disrupt the cell. Some of these techniques, such as bead milling, have already found applications on a large scale (Prasad, 2010).

In liquid shearing, the disruption takes place in a liquid medium and is caused by a sudden pressure drop, which happens when the suspended cells pass from high pressure into a chamber at atmospheric pressure through a small slit or orifice. This large pressure drop causes cavitations and shockwaves that disrupt the cell. The rate of cell disruption depends on the nature of the cells and their concentration (Prasad, 2010; Sivasankar, 2006). On the other hand, in solid shearing the cells are frozen at $-25°C$, and then pass through an orifice at high pressure. The disruption in this case happens due to the combination of liquid shearing through a narrow orifice and the solid shear of ice crystals (Panesar et al., 2010; Stanbury and Whitaker, 1984). Freezing and then thawing is another mechanical method, where ice crystals are formed due to freezing and then the expansion due to thawing leads to cell disruption. For efficient disruption, multiple cycles are usually required. However, repeated cycles of freezing and thawing may denature the enzymes produced.

In addition to these conventional methods, ultrasonication has also been used, where shock waves are created due to the generated high frequency vibrations that disrupt the cells. It is a simple method, but consumes a lot of power, uses probes that have short working times, and generates heat during the process. Thus, most ultrasonication systems are contained in a cooling jacket.

1.4.2 CHEMICAL AND ENZYMATIC CELL-DISRUPTION METHODS

Cells can also be disrupted chemically by detergents, osmotic shocks, organic solvents, and alkali treatments (Prasad, 2010). Detergents are molecules with hydrophilic and hydrophobic properties that allow them to interact both with water and lipids. The hydrophobic function is usually ionic, whereas the hydrophobic aspect is normally a hydrocarbon (Joesten et al., 2006; Marriott and Gravani, 2006; Tadros, 2005). They disrupt cell membranes by penetrating between the layers and forming micelles that separate lipids and proteins. Many detergents are available for cell membrane solubilization such as sodium dodecyl sulfate and triton X. However, many of them, such as sodium dodecyl sulfate, break protein–protein interaction and denature the enzyme.

A sudden change in salt concentration is referred to as osmotic shock. Cells are placed in a buffer solution at a high salt concentration and this results in water movement from inside the cell toward the buffer medium. The cells are then transferred into a weak buffer solution or distilled water. The sudden change in concentration results in cell disruption.

Organic solvents such acetone, chloroform, or methanol can also be used to extract enzymes. These chemicals liberate the enzyme by creating pores and channels in the cell membrane. However, organic solvents use is limited in large-scale production due to their toxicity, flammability, high costs, and possible protein denaturation effects (Ghosal and Srivastava, 2009).

As an alternative to chemical disruption, enzymes can be used. Enzymes have the advantage of operating at mild conditions and have comparatively low energy requirements. Lysozyme is an example of an enzyme that can be used for a large number of applications.

However, none of these methods is perfect. Therefore, a combination of two or more methods has been suggested in order to enhance disruption, save energy, and facilitate subsequent processing (D'souza and Killedar, 2008). There are two common combined approaches: (1) combining non-mechanical methods and (2) pretreating with chemical or enzymatic methods followed by mechanical methods.

In cell disruption, not only the enzymes but other cell contents, such as nucleic acids are also released, which need to be removed from the mixture either by precipitation or digestion. In addition, enzymes can be contaminated by undisrupted cells and cell debris that also need to be removed during purification. This can accomplished using centrifugation or filtration.

After the removal of the nucleic acid, cell debris, unwanted contaminants, and other proteins, water then needs to be removed. Different methods are available for enzyme concentration and initial purification. Among them are precipitation using ammonium sulfate, organic solvents or polymers, membrane separation, or an aqueous biphasic system (Janson, 2011; Schmauder and Schweizer, 1997).

Following initial purification, final purification via chromatography is required. This is well known for its high resolution and purification that removes impurities and brings the product close to final specifications. There are many chromatographic techniques such as size exclusion, ion exchange, hydrophobic interactions, and affinities that can be used for final enzyme purification. Each has its limitations that must be taken into consideration before selection.

Size exclusion chromatography, also known as gel filtration, is a technique that separates molecules based on their size or, more technically, their hydrodynamic particle volume (Ghosal and Srivastava, 2009). A stationary phase is a molecular sieve made from hydrophobic polymers that swells when water is absorbed and acts as a swelling gel. In such a technique, larger molecules pass through the column at a fast rate in the mobile phase, which is usually conducted in an aqueous solution. Small molecules are retained in the stationary phase and penetrate through the gel pores, depending on their size and shape. Elute is collected in a series of tubes based on the elution time. This technique is usually combined with other techniques that further separate molecules based on characteristics such as their charge and the affinity of certain compounds.

In ion-exchange chromatography, the separation is based on the charge. Columns are prepared either for an anion exchange or a cation exchange. Anion columns contain a stationary phase with a positive charge that attracts negatively charged proteins, whereas cation columns contain a negative charge that attracts positively charged proteins.

A hydrophobic interaction is based on the interaction between the hydrophobic ligands and hydrophobic areas located on the enzyme surface due to certain amino acid residues. Hydrophobic groups are attached to the stationary column. Enzymes that pass through the column and have hydrophobic residues on their surfaces are able to interact and bind.

Affinity chromatography relies on the ability of enzymes to bind to ligands that are bound to an inert support matrix packed in the column. Thus, only enzymes that selectively bind to ligands will be retained in the column. Unbound enzymes will be flushed out by column washing using an appropriate buffer solution that contains free ligands.

Generally, purification is measured by determining the enzyme specific gravity using a spectrophotometer, sodium dodecyl sulfate gel electrophoresis, isoelectric focusing, and a mass spectrophotometer.

1.5 INDUSTRIAL APPLICATIONS OF ENZYMES

Industrial enzymes can be divided into two major categories based on their application. These are technical and food enzymes. Technical enzymes include those used in detergents, textiles, leather, pulp, paper, and biofuels applications. The estimated global industrial enzyme market in the year 2000 was reported as approximately $1.5 billion (Shanmugam, 2009), where the largest proportion of market sales was for technical enzymes at 65%. Food enzymes were the second largest with 25% of the market. This includes enzymes used in brewing, wine and juice production, making fats and oils, and in the baking industries. Amylase, lipase, protease, lactase, laccase, phytase, cellulase, and xylanase are the common enzymes used in these industries. The majority of them are hydrolytic enzymes and used to degrade many natural substances (Galante and Formantici, 2003).

Protease is an enzyme used widely for detergents and dairy production, baking, and also in the pulp, paper, and leather industries. Carbohydrases, primarily amylases and cellulases, are used in starches, textiles, detergents, and the baking industry. Lipases are used in food, pulp, paper, leather, and the organic synthesis industries.

TABLE 1.4

Enzymes Used in Various Industrial Areas and Their Applications

Industry	Enzyme	Application
Detergent	Amylase	Removal of starch stains
	Protease	Removal of protein stains
	Lipase	Removal of oily and fatty stains
	Cellulase	Modification of the structure of cellulose fibers in cotton and cotton blends
Textile	Amylase	Desizing, denim finishing
	Cellulase	Polishing
	Catalase	Bleach termination
	Laccase	Bleaching
	Peroxidase	Removal of dyes
Leather	Lipase	Depickling
	Protease	Bating and unhearing
Pulp and paper	Amylase	Modification of starches
	Protease	Removal of biofilms
	Lipase	Pitch treatments
	Cellulase	Deinking
	Xylanase	Bleaching
	Laccase	Bleaching
Starch and fuel	Amylase	Starch liquefaction
	Glucose isomerase	Starch sweetening
	Xylanase	Fuel viscosity reduction
Food	Amylase	Bread softness and volume, flour adjustment
	Lipase	Cheese flavor formation
	Protease	Infant formulas, flavor
	Lactase	Lactose removal
	Lysozyme	Low bowling prevention in chesses making
	Transglutaminase	Changing dairy products rheological properties and textural improvements
	Glucose oxidase	Raw milk preservation

Novozymes A/S in Denmark, Genencor International Inc. in the United States, and DSM N.V. in the Netherlands are the three leading industrial enzymes suppliers. Examples of the industrial applications of enzymes are given in Table 1.4.

1.5.1 Detergents

Enzymes are used as ingredients in laundry and dishwashing detergents to improve their efficiency. Their functional efficiency depends on the detergent's components, type of stain to be removed, water temperature, and hardness. Major enzymes used for this purpose are proteases, amylases, lipases, and cellulases. Proteases, which are the most widely used enzymes in the detergent industry, enhance the breakdown of protein constituents through the hydrolysis of amide bonds between amino acids into more soluble polypeptides or free amino acids. Amylases facilitate the removal of starchy residues from foodstuffs, such as potatoes and spaghetti, by catalyzing the hydrolysis of glycosidic linkages in starch polymers; α-amylases are the most commonly used. Lipases remove butter, oil, chocolate, and cosmetic stains, whereas cellulases modify the structure of cellulose fibers in cotton and cotton blends by removing microfibrils from cotton and cotton-blended fabrics.

1.5.2 TEXTILES

In the textile industry many steps are necessary. These include yarn manufacturing, where the fiber is converted to yarn and then into fabric. Fabrics are then dyed and printed to make the finished product. Enzymes like amylase, pectinase, catalase, and cellulase are commonly used in applications such as desizing, bioscouring, bleaching, dying, and biopolishing (Galante and Formantici, 2003). Figure 1.4 shows the major steps in cotton textile manufacturing and each enzyme used at each stage.

In fabric manufacturing, threads are coated prior to weaving with a substance that prevents threads breaking during weaving. Many substances such as starch and polyvinyl alcohol have been used for this purpose. However, these should be removed after weaving. The desizing (size removal) process is usually carried out using chemicals such as acids, alkyls, or oxidizing agents. However, these chemicals are not recommended due to their environmental impact. Starch-breaking enzymes (amylase) are used instead due to their high efficiency in removing size without affecting the fabric and its specific character.

One of the most energy-consuming (and water-consuming) steps in the textile industry is the scouring stage. This is the removal of noncellulosic compounds from cellulose fibers at high temperatures and in strong alkaline conditions (Kirk et al., 2002). For this, alkaline chemicals are used. However, these chemicals do not only remove noncellulosic components but also attack the cellulose, and that causes a loss in fabric weight and strength. As an alternative, enzymes such as pectinase have been used in milder conditions.

Bleaching is another step where enzymes, such as catalase, are applied. Generally, before dying, fabrics are bleached with hydrogen peroxide. The remaining hydrogen peroxide after the bleaching reaction has finished may obstruct the dying process. Traditionally, this is solved by using a reducing agent. This requires a controlled amount of the agent. Thus, using enzymes is a suitable alternative, where a small amount can break down hydrogen peroxide into water and oxygen.

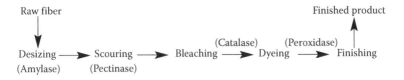

FIGURE 1.4 Enzymes used in different textile manufacturing steps.

Cellulases are another important group of enzymes used in the textile industry due to their ability to modify cellulosic fibers and improve their quality.

1.5.3 LEATHER

Hide and skin processing for leather production involves several steps. These are socking, dehairing, bating, and tanning. Arunachalam and Saritha (2009) reported that hides and skins have a chemical composition of 60% to 65% water, 25% to 30% protein, and 5% to 10% fat with traces of minerals.

In the hair removal process the usual technique is to use sodium sulfide (Covington and Covington, 2009), which swells the hide and destroys the hair. Although this method is cheap, it is environmentally unfriendly (Arunachalam and Saritha, 2009; Covington and Covington, 2009). This pushed the leather industry to use proteases instead. Alkaline proteases, which are usually derived from *Bacillus* strains, have been successfully used to speed up this step (Rao et al., 1998). After dehairing and prior to tanning, hides have to be softened by a process called bating, which uses enzymes (Galante and Formantici, 2003). In this step, proteins are broken down and removed using proteases and trypsin (Rao et al., 1998).

1.5.4 PULP AND PAPER

Traditionally, the pulp and paper industries used chemicals at various stages. In order to reduce the environmental impact of the chemicals, they were replaced by enzymes. Jaworski (2001) documented that enzymes used in the pulp and paper industries reduces the use of chlorine by 10% to 15% and water by 18%. In addition, the use of enzymes improves paper brightness and strength.

Proteases, amylases, cellulases, xylenases, esterases, and lipases are the main enzymes used in this industry. Amylases are used for starch removal, cellulases for improving strength, xylenases for xylene removal from the pulp, increasing pulp strength and providing more brightness. Esterase is used for stick removal and lipases for controlling the pitch.

1.5.5 BIOFUEL PRODUCTION

1.5.5.1 Biodiesel

Biodiesel, which is a fatty acid methyl ester, is a good replacement for petroleum diesel, because it is renewable, biodegradable, and less toxic. Commercially, biodiesel is produced using an alkali process. This is a cost-effective and highly efficient process. However, to avoid downstream process problems, the feed must be purified and free of fatty acids.

Biodiesel production using the biocatalyst lipase avoids the disadvantages of the alkaline process without the need for subsequent wastewater treatment (Taher et al., 2011). Additionally, lipases can operate in mild conditions with a high substrate selectivity. *M. miehei*, *R. oryzae*, *C. antarctica*, and *P. cepacia* are all common lipases found to be capable of catalyzing oil transesterification in order to produce biodiesel.

1.5.5.2   Bioethanol

Methyl tertiary-butyl ether (MTBE) is added to gasoline as an oxygenate to raise the oxygen content of gasoline. This prevents the engine from knocking. But MTBE is a volatile and flammable liquid that dissolves easily in water at room temperature. Therefore, the U.S. Environmental Protection Agency has announced regulatory action to eliminate MTBE from gasoline (Hamid and Ali, 2004). Bioethanol, on the other hand, is a better alternative to MTBE. Current production of bioethanol relies on microbial fermentation of starches and sugars. However, there have been considerable arguments about its sustainability, as its feedstock competes for food sources and its energy costs are relatively high compared to fossil fuels. A possible source for low-cost production is to utilize lignocellulosic material such as crop residues, grasses, sawdust, wood chips, and solid animal waste (Sun and Cheng, 2002).

The conversion of lignocellulosic materials to bioethanol has two steps. They are (1) the hydrolysis of cellulose in the lignocellulosic material to fermentable reducing sugars and (2) the fermentation of the sugars into ethanol. Hydrolysis is usually catalyzed by a cellulase enzyme, and the fermentation is carried out by yeast or bacteria. A factor affecting cellulose hydrolysis is its lignin and hemicellulose content. The presence of lignin and hemicellulose makes access to cellulose difficult and thus reduces the efficiency of the hydrolysis.

The cellulase enzymes used for the hydrolysis of cellulose are highly specific (Wongskeo et al., 2012). Cellulases are usually a mixture of several enzymes. At least three major groups of cellulases are involved in the hydrolysis process: (1) endoglucanase (EC 3.2.1.4), which attacks regions of low crystallinity in the cellulose fiber thus creating free chain-ends; (2) exoglucanase or cellobiohydrolase (EC 3.2.1.91), which degrades the molecule further by removing cellobiose units from the free chain-ends; and (3) β-glucosidase (EC 3.2.1.21), which hydrolyzes cellobiose to produce glucose (Samsuri et al., 2009).

A large range of microorganisms can produce a complete set of cellulases. *Trichoderma reesei* is a well-known fungus that can produce cellobiohydrolases, endoglucanases, and β-glucosidases, which are all necessary to efficiently hydrolyze cellulose (Ferreira et al., 2009). In addition, *T. reesei* is able to produce hemicellulases, mainly xylanases, depending on the growth conditions and substrate. It is well known that the conjugated action of cellulases and hemicellulase results in higher final sugar production compared to cellulase alone.

1.5.6   FOOD INDUSTRY

Enzymes have several applications in the food industry, including baking, cheese making, in coffee and tea production and making fruit juices and wines. Enzymes in these applications must be obtained from sources that are generally regarded as safe (Middelberg, 1995). Thus, the majority of enzymes come from sources that are already employed in the food industry.

1.5.6.1   Bakery Industry

The flour component that affects a dough's rheological properties is the gluten protein. This plays a key role in providing the required quality of wheat by producing water absorption capacity, viscosity, elasticity, and cohesiveness. Generally, the quality of bread making is determined by the quality of the gluten. Gluten proteins consist of monomeric gliadins, connected by disulfide bonds (Wieser, 2007). During mixing and molding the dough, SS bonds are formed and broken between gluten proteins. When this take place, it leads to good dough development, and strong gluten networks are formed. Usually, chemical oxidants such as bromide are used to support bond formation and improve the flour. Nevertheless, it has been reported that adding bromate can have hazardous effects, thus it has been forbidden in many countries (Corrales et al., 1993). To find a substitute for bromate, ascorbic acid, commonly known as vitamin C, was considered and used as a flour improver in spite of it being less effective than bromate (Ranum, 1992). Due to concerns about using chemical oxidizing agents, the food industry searched for natural alternatives with a higher specificity. Enzymes that denaturate during baking have been proposed. Oxidoreductases have a beneficial effect on dough development and dough quality, where the latter influences other parameters such as volume, texture, and the crumb structure of baked products (Joye et al., 2009; Vemulapalli et al., 1998).

1.5.6.2 Dairy Industry

The use of enzymes in making dairy products is one of their oldest applications. Some enzymes are used for cheese, yogurt, and other dairy products, whereas others are used for texture and flavor improvements. Generally, proteases are added to milk during cheese production to hydrolyze caseins. In the presence of molecular oxygen, glucose oxidase catalyzes the oxidation of B-d-glucose to d-gluconic acid and hydrogen peroxide that can serve as an alternative to calcium peroxide (Liao et al., 1998). Hexose oxidase is another enzyme that has the ability to oxidize oligosaccharides. Similarly to glucose oxidase, it can generate SS bonds between proteins due to hydrogen peroxide formation (Poulsen and Hostrup, 1998).

Sweeteners used throughout the world are usually derived from starch. Frequently, these are produced by an acid hydrolysis process into simpler carbohydrates. Nowadays, acid hydrolysis has been replaced by enzymes. In the enzymatic process, the treatment of starch results in various syrups that are used in the food, beverage, and pharmaceutical industries.

Fructose syrup production has three steps. Amylase is first added to liberate maltodextrins that contain dextrins and oligosaccharides. In the second step, dextrins and oligosaccharides are completely hydrolyzed by pullulanase and glucoamylase, through a saccharification process, to glucose, maltose, and isomaltose. This is followed by the third step where glucose isomerase is used to treat glucose/maltose and convert glucose to fructose (Prasad, 2010).

1.6 ENZYME ENGINEERING

Enzymes work effectively in environments similar to those where they were originally extracted. Although large numbers of isolated enzymes have received attention, many of them do not fulfill the specific needs that they have to meet.

Many industrial processes require enzymes to operate at conditions in which they become unstable or inactive and not suitable without further modification. Thus, enzyme engineering improves enzyme functionality. Altering enzyme enantioselectivity (Carr et al., 2003; Otten et al., 2010) and enhancing their stability in severe conditions, such as extreme temperatures and pH or with aggressive chemicals (Illanes, 1999; Lozano et al., 1996; Merz et al., 2000), are the main improvements.

In protein engineering, enzyme properties are altered by changing the structure, which can be performed chemically by modifying the protein amino acid residues or by altering their primary structure. Normally, chemical modification of enzymes is carried out by a cross-linking of the desired amino acid. It is possible to alter the optimum pH, enzyme reactivity, and the type of reaction that can be catalyzed (DeSantis and Jones, 1999; Kaiser et al., 1985). However, using chemicals requires extracting the protein from the cell and sometimes purifying it. Therefore it is not applicable for proteins that should be active intracellularly. Thus, nowadays, it is more promising to directly change the properties of enzymes by altering the primary structure.

Genetic engineering means it is possible to introduce some genes or foreign DNA into the organism of interest. This is referred to as recombined DNA technology. This is accomplished by isolating the gene and copying it to generate a DNA sequence that contains the required genetic element, and then inserting it into the organism. By doing this, the desired enzymes can be produced. Lipolase, a detergent lipase produced by Novo Nordisk A/S, is an example of a genetically modified lipase that improves fat stain removal (Estell 1993, Smith, 2004).

REFERENCES

Arunachalam, C., and K. Saritha. 2009. "Protease Enzyme: An Eco-Friendly Alternative for Leather Industry." *Indian Journal of Science and Technology* 12 (12):29–32.

Balkan, B., and F. Ertan. 2005. "Production and Properties of Alpha-Amylase from Penicillium Chrysogenum and Its Application in Starch Hydrolysis." *Preparative Biochemistry and Biotechnology* 35 (2):169–178.

Bartlett, P. N., and R. G. Whitaker. 1987. "Strategies for the Development of Amperometric Enzyme Electrodes." *Biosensors* 3 (6):359–379.

Braverman, J. B. S., and Z. Berk. 1976. *Introduction to the Biochemistry of Foods*. Elsevier Scientific.

Campbell, M. K., and S. O. Farrell. 2011. *Biochemistry*. 7th ed. Brooks Cole.

Carr, R., M. Alexeeva, A. Enright, T. S. C. Eve, M. J. Dawson, and N. J. Turner. 2003. "Directed Evolution of an Amine Oxidase Possessing Both Broad Substrate Specificity and High Enantioselectivity." *Angewandte Chemie International Edition* 42 (39):4807–4810.

Chang, C. T., M. S. Tang, and C. F. Lin. 1995. "Purification and Properties of Alpha-Amylase from Aspergillus Oryzae Atcc 76080." *Biochemistry & Molecular Biology International* 36 (1):185–193.

Chisti, Y., and M. Moo-Young. 1986. "Disruption of Microbial Cells for Intracellular Products." *Enzyme and Microbial Technology* 8:194–204.

Constantinides, A., W. R. Vieth, and P. M. Fernandes. 1973. "Characterization of Glucose Oxidase Immobilized on Collagen." *Molecular and Cellular Biochemistry* 11 (1):127–133.

Corrales, X., M. Guerra, M. Granito, and J. Ferris. 1993. "Substitution of Ascorbic Acid for Potassium Bromide in the Making of French Bread." *Archivos Latinoamericanos de Nutrición* 24 (3):234–240.

Costa, S. A., H. S. Azevedo, and R. L. Reis. 2005. "Enzyme Immobilization in Biodegradable Polymers for Biomedical Applications." In *Biodegradable Systems in Tissue Engineering and Regenerative Medicine*, edited by R. L. Reis and J. San Román, 301–323. CRC Press.

Covington, A., and T. Covington. 2009. *Tanning Chemistry: The Science of Leather*. Royal Society of Chemistry.

Cygler, M., and J. D. Schrag. 1999. "Structure and Conformational Flexibility of Candida Rugosa Lipase1." *Biochimica et Biophysica Acta (BBA)—Molecular and Cell Biology of Lipids* 1441 (2–3):205–214.

D'souza, J. I., and S. G. Killedar. 2008. *Biotechnology and Fermentation Process*. Nirali Prakashan.

de Souza, P. M., and P. de Oliveira e Magalhães. 2010. "Application of Microbial alpha-Amylase in Industry: A Review." *Brazilian Journal of Microbiology* 41:850–861.

Deng, Y., and S. Fong. 2010. "Influence of Culture Aeration on the Cellulase Activity of *Thermobifida Fusca*." *Applied Microbiology and Biotechnology* 85 (4):965–974.

DeSantis, G., and J. B. Jones. 1999. "Chemical Modification of Enzymes for Enhanced Functionality." *Current Opinion in Biotechnology* 10 (4):324–330.

Estell, D. A. 1993. "Engineering Enzymes for Improved Performance in Industrial Applications." *Journal of Biotechnology* 28 (1):25–30.

Ferreira, S. M. P., A. P. Duarte, J. A. Queiroz, and F. C. Domingues. 2009. "Influence of Buffer Systems on Trichoderma Reesei Rut C-30 Morphology and Cellulase Production." *Electronic Journal of Biotechnology* 12 (3).

Finkelstein, A. E., E. S. Strawich, and S. Sonnino. 1970. "Characterization and Partial Purification of a Lipase from *Pseudomonas Aeruginosa*." *Biochimica et Biophysica Acta (BBA)—Enzymology* 206 (3):380–391.

Fish, N. M., and M. D. Lilly. 1984. "The Interactions between Fermentation and Protein Recovery." *Nature Biotechnology* 2 (7):623–627.

Freifelder, D., and G. M. Malacinski. 1993. *Essentials of Molecular Biology*. Jones and Bartlett.

Galante, Y. M., and C. Formantici. 2003. "Enzyme Applications in Detergency and in Manufacturing Industries." *Current Organic Chemistry* 7 (13):1399–1422.

Geciova, J., D. Bury, and P. Jelen. 2002. "Methods for Disruption of Microbial Cells for Potential Use in the Dairy Industry—A Review." *International Dairy Journal* 12 (6):541–553.

Ghosal, S., and A. K Srivastava. 2009. *Fundamentals of Bioanalytical Techniques and Instrumentation*. Prentice-Hall of India.

Godfrey, T., and S. West. 1996. *Industrial Enzymology*. Stockton Press.

Goss, J. A. 1973. *Physiology of Plants and Their Cells*. Pergamon Press.

Gouda, M., and Y. Elbahloul. 2008. "Statistical Optimization and Partial Characterization of Amylases Produced by Halotolerant *Penicillium* Sp." *World Journal of Agricultural Sciences* 4 (3):359–368.

Hamid, H., and M. A. Ali. 2004. *Handbook of MTBE and Other Petroleum Oxygenates*. Marcel Dekker.

Hatti-Kaul, R., and B. Mattiasson. 2003. *Isolation and Purification of Proteins*. Marcel Dekker.

Hermida, C., G. Corrales, F. J. Cañada, J. J. Aragón, and A. Fernández-Mayoralas. 2007. "Optimizing the Enzymatic Synthesis of beta-D-Galactopyranosyl-D-Xyloses for Their Use in the Evaluation of Lactase Activity in Vivo." *Bioorganic and Medicinal Chemistry* 15 (14):4836–4840.

Hirata, A., M. Adachi, S. Utsumi, and B. Mikami. 2004. "Engineering of the pH Optimum of Bacillus Cereus Beta-Amylase: Conversion of the pH Optimum from a Bacterial Type to a Higher-Plant Type." *Biochemistry* 43 (39):12523–12531.

Illanes, A. 1999. "Stability of Biocatalysts." *Electronic Journal of Biotechnology* 2 (1):7–15.

Janson, J. C. 2011. *Protein Purification: Principles, High Resolution Methods, and Applications*. 3rd ed. John Wiley & Sons.

Jaworski, J. 2001. *The Application of Biotechnology to Industrial Sustainability: A Primer*. Canada.

Joesten, M. D., J. L. Hogg, and M. E. Castellion. 2006. *The World of Chemistry: Essentials*. Thomson Brooks/Cole.

Joye, I. J., B. Lagrain, and J. A. Delcour. 2009. "Use of Chemical Redox Agents and Exogenous Enzymes to Modify the Protein Network During Breadmaking: A Review." *Journal of Cereal Science* 50 (1):11–21.

Kaiser, E. T., D. S. Lawrence, and S. E. Rokita. 1985. "The Chemical Modification of Enzymatic Specificity." *Annual Review of Biochemistry* 54:565–595.

Kalisz, H. M., J. Hendle, and R. D. Schmid. 1997. "Structural and Biochemical Properties of Glycosylated and Deglycosylated Glucose Oxidase from *Penicillium Amagasakiense*." *Applied Microbiology and Biotechnology* 47 (5):502–507.

Kirk, O., T. V. Borchert, and C. C. Fuglsang. 2002. "Industrial Enzyme Applications." *Current Opinion in Biotechnology* 13 (4):345–351.

Kona, R. P., N. Qureshi, and J. S. Pai. 2001. "Production of Glucose Oxidase Using Aspergillus Niger and Corn Steep Liquor." *Bioresource Technology* 78 (2):123–126.

Lee, S., Y. Mouri, M. Minoda, H. Oneda, and K. Inouye. 2006. "Comparison of the Wild-Type alpha-Amylase and Its Variant Enzymes in Bacillus Amyloliquefaciens in Activity and Thermal Stability, and Insights into Engineering the Thermal Stability of Bacillus alpha-Amylase." *Journal of Biochemistry* 139 (6):1007–1015.

Liao, Y., R. A. Miller, and R. C. Hoseney. 1998. "Role of Hydrogen Peroxide Produced by Baker's Yeast on Dough Rheology1." *Cereal Chemistry Journal* 75 (5):612–616.

Lorenz, P., and J. Eck. 2005. "Metagenomics and Industrial Applications." *Nature Reviews Microbiology* 3 (6):510–516.

Lozano, P., A. Avellaneda, R. Pascual, and J. L. Iborra. 1996. "Stability of Immobilized alpha-Chymotrypsin in Supercritical Carbon Dioxide." *Biotechnology Letters* 18 (11):1345–1350.

Margesin, R., F. Schinner, J. C. Marx, and C. Gerday. 2008. *Psychrophiles: From Biodiversity to Biotechnology*. Springer.

Marriott, N. G., and R. B. Gravani. 2006. *Principles of Food Sanitation*. Springer.

Merz, A., M.-C. Yee, H. Szadkowski, G. Pappenberger, A. Crameri, W. P. C. Stemmer, C. Yanofsky, and K. Kirschner. 2000. "Improving the Catalytic Activity of a Thermophilic Enzyme at Low Temperatures." *Biochemistry* 39 (5):880–889.

Middelberg, A. P. J. 1995. "Process-Scale Disruption of Microorganisms." *Biotechnology Advances* 13 (3):491–551.

Otten, L. G., F. Hollmann, and I. W. C. E. Arends. 2010. "Enzyme Engineering for Enantioselectivity: From Trial-and-Error to Rational Design?" *Trends in Biotechnology* 28 (1):46–54.

Panesar, P. S., S. S. Marwaha, and H. K. Chopra. 2010. *Enzymes in Food Processing: Fundamentals and Potential Applications*. I.K. International Publishing House.

Pilarek, M., and K. W. Szewczyk. 2007. "Kinetic Model of 1,3-Specific Triacylglycerols Alcoholysis Catalyzed by Lipases." *Journal of Biotechnology* 127 (4):736–744.

Poulsen, C., and P. B. Hostrup. 1998. "Purification and Characterization of a Hexose Oxidase with Excellent Strengthening Effects in Bread." *Cereal Chemistry Journal* 75 (1):51–57.

Prasad, K. 2010. *Downstream Process Technology: A New Horizon in Biotechnology*. Prentice-Hall of India.

Ranum, P. 1992. "Potassium Bromate in Bread Baking." *Cereal Foods World* 37 (3):261–263.

Rao, M. B., A. M. Tanksale, M. S. Ghatge, and V. V. Deshpande. 1998. "Molecular and Biotechnological Aspects of Microbial Proteases." *Microbiology and Molecular Biology Reviews* 62 (3):597–635.

Ray, R. R. 2000. "Purification and Characterization of Extracellular Beta-Amylase of Bacillus Megaterium B(6)." *Acta Microbiol Immunol Hung* 47 (1):29–40.

Samsuri, M., M. Gozan, B. Prasetya, and M. Nasikin. 2009. "Enzymatic Hydrolysis of Lignocellulosic Bagasse for Bioethanol Production." *Journal of Biotechnology Reseaech in Tropical Region* 2 (2):1–5.

Schmauder, H. P., and M. Schweizer. 1997. *Methods in Biotechnology*. Taylor & Francis.
Shanmugam, S. 2009. *Enzyme Technology*. I.K. International Publishing House.
Sivasankar, B. 2006. *Bioseparations: Principles and Techniques*. Prentice-Hall of India.
Smith, J. E. 2004. *Biotechnology*. Cambridge University Press.
Spencer, J. N., G. M. Bodner, and L. H. Rickard. 2010. *Chemistry: Structure and Dynamics*. 5th ed. Wiley.
Stanbury, P. F., and A. Whitaker. 1984. *Principles of Fermentation Technology*. Pergamon Press.
Steward, F. C., and R. G. S. Bidwell. 1991. *Plant Physiology: A Treatise*. Academic Press.
Stred'anský, M., M. Tomáška, E. Šturdík, and L. Kremnický. 1993. "Optimization of beta-Galactosidase Extraction from *Kluyveromyces Marxianus*." *Enzyme and Microbial Technology* 15 (12):1063–1065.
Sun, Y., and J. Cheng. 2002. "Hydrolysis of Lignocellulosic Materials for Ethanol Production: A Review." *Bioresource Technology* 83 (1):1–11.
Szajáni, B., A. Molnár, G. Klámar, and M. Kálmán. 1987. "Preparation, Characterization, and Potential Application of an Immobilized Glucose Oxidase." *Applied Biochemistry and Biotechnology* 14 (1):37–47.
Tadros, T. F. 2005. *Applied Surfactants: Principles and Applications*. Wiley-VCH.
Taher, H., S. Al-Zuhair, A. H. Al-Marzouqi, Y. Haik, and M. Farid. 2011. "A Review of Enzymatic Transesterification of Microalgal Oil-Based Biodiesel Using Supercritical Technology." *Enzyme Research* 2011 (Article ID 468292).
Takasaki, Y. 1983. "An Amylase Producing Maltotetraose and Maltopentaose from *Bacillus Circulans*." *Agricultural and Biological Chemistry* 47:2193–2199.
Ul-Haq, I., H. Ashraf, and S. Ali. 2007. "Kinetic Characterization of Extracellular Alpha-Amylase from a Derepressed Mutant of Bacillus Licheniformis." *Applied Biochemistry and Biotechnology* 1 (3):251–164.
Vemulapalli, V., K. A. Miller, and R. C. Hoseney. 1998. "Glucose Oxidase in Breadmaking Systems." *Cereal Chemistry* 75 (4):439–442.
Wieser, H. 2007. "Chemistry of Gluten Proteins." *Food Microbiology* 24 (2):115–119.
Wongskeo, P., P. Rangsunvigit, and S. Chavadej. 2012. "Production of Glucose from the Hydrolysis of Cassava Residue Using Bacteria Isolates from Thai Higher Termites." *World Academy of Science, Engineering and Technology* 64:353–365.

2 Lipases

Lipases are hydrolytic enzymes that have received extensive attention in food, pulp and paper and fuels industries. They can be found in microorganisms, plants, and animals with ability to breakdown of lipids. In this chapter, lipases are introduced and there is a brief discussion about their functions. This is followed by a short description of their main sources, structure, and features with an emphasis on their specificity and interfacial activity. The chapter focuses on microbial lipases, which are usually preferred for commercial applications due to their favorable properties, easy extraction, and unlimited supply. The chapter concludes with a discussion on the various industrial applications of lipases and properties improvement.

2.1 LIPASES, ESTERASES, AND PHOSPHOLIPASES

Lipids are large, water-insoluble molecules that composed of triglycerides of fatty acids. They play important biological roles in conjunction with lipolytic enzymes. These enzymes are esterases (EC 3.1.1.1) and lipases (EC 3.1.1.3), which differ in their specificity toward the substrates. Esterases act on short chains of triglycerides (water soluble), whereas lipases act on lipids (water insoluble) (Chahinian and Sarda, 2009; Fojan et al., 2000).

Lipases are water-soluble enzymes that naturally catalyze the hydrolysis of oils and fats, insoluble substrates, acting at the oil–water interface. Their optimum activity of lipases is reached in emulsions, where high substrate surfaces can be obtained and substrate concentrations exceed the critical micelle concentration (Meyer et al., 1990).

The International Union of Biochemistry and Molecular Biology (IUBMB) defines lipases as glycerol-ester hydrolases and recommends that ester emulsion should be used as the substrate. The first lipase use was investigated by Claude Bernard in 1856. He studied the role of pancreas juice in fat digestion (Petersen and Drabløs, 1994). This is a distinct feature of lipases, commonly referred to as an interface activation. Besides the interface action, lipases can also work in bulk solutions. That, however, is at lower activity levels compared to those at the interfaces (Reetz, 2002; Taipa et al., 1992; Weete et al., 2008). The kinetics of a lipase-activated reaction, therefore, are described by interfacial activation kinetics rather than the well-known Michaelis-Menten kinetics (Sarda and Desnuelle, 1958). This phenomenon is further explained in Section 2.3.3.

In addition to hydrolysis, lipases can also catalyze transesterification and interesterifications. This diversity allows lipases to be used for various applications in the field of chemical synthesis, including oil and fat splitting and modification. Details of lipase use in various applications are provided in Section 2.6.

Lipases differ from phospholipases. By the former, both triacylglycerols and phospholipidsare are hydrolyzed, whereas by the presence of the latter only phospholipids

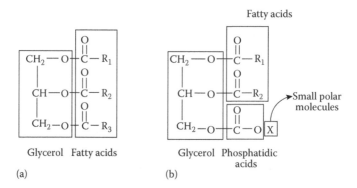

FIGURE 2.1   Chemical structure of (a) triacylglycerols and (b) phospholipids.

are hydrolyzed (Ghosh et al., 2005). Phospholipids are similar to triacylglycerols, but one of the three fatty acids has been replaced by a phosphate functional group, as shown in Figure 2.1. The focus of this chapter is on lipases that can hydrolyze triacylglycerols.

2.2   SOURCES OF LIPASES

Lipases can be extracted from microbial organisms (bacteria, yeast, and fungi), plants (wheat, oats, corns, and palms), or animals (pancreas, stomach, pharynx, and other tissues) (Taipa et al., 1992). Regardless of the source, most lipases have similar three-dimensional structures and are able to catalyze similar reactions. Nevertheless, they may differ under the same reaction conditions (Yahya et al., 1998). However, Jaeger et al. (1994) showed that there are large differences between amino acid sequences of screened lipases obtained from different sources.

Lipases from several mammalian species have been investigated, and pancreatic lipases were studied in more detail. Plant lipases have not received as much attention as mammalian, and their applications are more in lipid biotransformation (Mukherjee, 1994). Generally, commercially available industrial lipases are derived from microorganisms (Hasan et al., 2006), especially from those that produce extracellular lipases (Sharma et al., 2001). This is mainly due to their availability and versatility as well as their rapid growth in inexpensive media. Furthermore, the production of microbial lipases is not seasonal, as is the case with plant-derived lipases. Genetic modifications of microbial lipases are also easier to engineer, which can result in more thermally stable and chemical-denaturation-resistant lipases (Hasan et al., 2006; Li and Zhang, 2005; Salleh et al., 2006; Yahya et al., 1998). The amount of lipase that can be produced from microorganisms depends on several factors, including culture temperature, pH, lipids, nutrient composition, and the availability of carbon and oxygen sources (Taipa et al., 1992). A lot of research has been carried out to investigate the influence of these factors on lipase production. Table 2.1 summarizes the properties of some of the most common lipases from different sources. Lipases from genera *Rhizomucor, Rhizopus, Thermomyces, Pseudomonas,* and *Candida* are the main lipases used commercially. Examples of such commercially available

TABLE 2.1
Properties of Some Characterized Microbial Lipases

Strain (Organism)	Molecular Weight (kDa)	Specific Activity (U/mg)	Optimum Conditions		References
			pH	Temperature (°C)	
Neurospora sp.	55	8203	6.5	45	Lin et al., 1996
Botrytis cinerea	60	2574	6	38	Comménil et al., 1995
Pseudomonas fluorescens	33	6200	8–10	55	Kojima et al., 1994
Rhizopus niveus	34	4966	6–6.5	35	Kojima et al., 1994
Neurospora crassa	54	44	7	30	Kundu et al., 1987
Rhizopus delemar	30	7638	8–8.5	30	Haas et al., 1992
Fusarium heterosporium	31	2010	5.5–6	45–50	Shimada et al., 1993
Penicikkium roquefortii	25	30	6–7	30	Tamio et al., 1995
Penicillium sp.	27	25	7	25	Gulomova et al., 1996
Acinetobacter calcoaceticus	23	–	7	50	Pratuangdejkul and Dharmsthiti, 2000
Bacillus subtilis 168	19	–	12	40	Lesuisse et al., 1993
Pseudomonas aeruginosa EF2	29	–	6.5–7.5	45–50	Gilbert, Cornish et al., 1991

lipases are Lipozyme RMM (immobilized form) from *Rhizomucor miehei*; Ta-lipase from *Rhizopus delemar*; Lipase PS from *Pseudomonas cepacia*; Lipozyme TLIM from *Thermomyces lanuginosus*; Novozym®435 from *C. antarctica*; and Lipase-OF from *C. rugosa*. Typically, most of these microbial lipases have a molecular weight of 19 to 40 kDa, however, *Geotrichum candidum* and *C. rugosa* show larger molecular weights of around 60 kDa (Holmquist, 2000).

2.3 LIPASE STRUCTURE AND INTERFACIAL ACTIVATION

2.3.1 LIPASE STRUCTURE

Several lipase structures have been identified by x-ray crystallography and nuclear magnetic resonance. The three-dimensional structures were first defined for human pancreatic lipase (Winkler et al., 1990) and the *Rhizomucor miehei* lipase (Brady et al., 1990). This was followed by the determination of the structures of the *C. rugosa* (Grochulski et al., 1993) and the *Pseudomonas glumae* (Noble et al., 1993) lipases. These structures have all suggested that lipases have an α/β fold. The active site for α/β hydrolase fold enzymes has three catalytic residues—a nucleophilic residue (ser-ine, cysteine, or aspartate), a histidine residue, and a catalytic acid residue (aspartate or glutamate)—in the amino acid sequence. In lipases, the nucleophilic amino acid has been found to be a serine residue (Ser), whereas the catalytic acid can either be an aspartate (Asp) or a glutamate (Glu) residue, as described by Ollis et al. (1992). The lipase core is composed of a central β sheet that consists of up to eight differ-ent β strands (β1–β8) connected by up to six α helices (A–F), as shown in Figure 2.2.

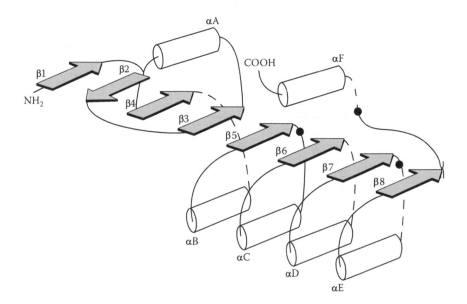

FIGURE 2.2 α/β Hydrolysis fold. (From Nardini, M., and B. W. Dijkstra, 1999, *Current Opinion in Structural Biology* 9 (6):732–737. With permission.)

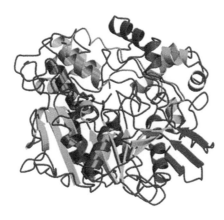

FIGURE 2.3 **(See color insert.)** Structure of *Candida rugosa* lipase. (From Cygler, M., and J. D. Schrag, 1999, *Biochimica et Biophysica Acta (BBA)—Molecular and Cell Biology of Lipids* 1441 (2–3):205–214. With permission.)

The α/β fold type is also common in other lipolytic enzymes, such as proteases (Ollis et al., 1992). Serine is a residue in the active structure of Ser-His-Asp, with a polypeptide flap (often called a lid) that covers the active sites and shields them and controls substrate access. The Ser-His-Asp structure is confirmed in all the analyzed lipases structures except for *G. candidum* lipase that has a structure of Ser-His-Glu (Jacobsen et al., 1990; Mala and Takeuchi, 2008). An example of a three dimensional structure for lipase is shown in Figure 2.3. The core is formed from parallel β-strands (central sheets are the light blue region and N-terminal sheets are the dark blue region in Figure 2.3) connected by a number of α-helices (the dark green region in Figure 2.3) that flank the sheet on both sides. The minimal fragment of this structure folds five β-sheets and two α-helices.

Lipase structures can be divided into two categories based on active site accessibility to the substrate, namely, (1) the open form, where the active sites are accessible to the solvent, and (2) the closed form, where the active sites are inaccessible. This is illustrated in Figure 2.4. However, both forms have been observed for certain lipases such as lipase from *P. cepacia* (Schrag et al., 1997). Mainly, the conformational state of the lipase depends on crystallization conditions. For example, polyethylene glycol conditions promote the closed form, whereas detergents and large alcohols promote an open form (Jaeger and Reetz, 1998). The presence of an inhibitor bound in an active site is another factor that affects the conformational state, where only the open confirmation can be observed (Miled et al., 2000; Salameh and Wiegel, 2010; Schrag et al., 1997).

2.3.2 Mechanism of Lipase Action

Due to the similarity between the lipase and protease catalytic triads, the mechanism of lipase catalysis is similar to that of serine protease (Jaeger et al., 1999). α/β Hydrolases have a common catalytic mechanism of ester hydrolysis. First, serine is activated by deprotonation. Subsequently, the nucleophility of hydroxyl

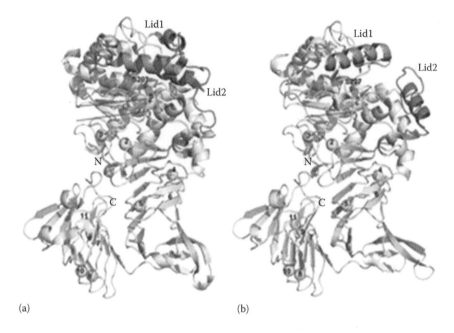

(a)   (b)

FIGURE 2.4   **(See color insert.)** Structure of *Pseudomonas* lipase in (a) closed and (b) open conformation. (From Cheng, M., C. Angkawidjaja, Y. Koga, and S. Kanaya, 2014, *Protein Engineering Design and Selection* 27 (5):169–176. With permission.)

residue is enhanced and attacks the substrate carbonyl group, forming a tetrahedral intermediate that releases an alcohol molecule to produce an acyl-enzyme inter-mediate. The presence of an oxyanion hole contributes to stabilizing the charge distribution. This is followed by a water molecule attack on the complex giving a tetrahedral intermediate that loses an acid molecule to produce the enzyme in its native form (Reis et al., 2009). Figure 2.5 shows the stages of the reaction cata-lyzed by lipase.

2.3.3   INTERFACIAL ACTIVATION

Normally, reactions take place at the oil–water interface by the movement of the lid allowing the substrate to access the active sites. This is a special phenomenon in lipases and is referred to as the interfacial activation. This distinguishes lipases from esterases. This phenomenon is often associated with the reorientation of the α-helical lid structure that in turn increases the surface hydrophobicity in the vicinity of the active site and subsequently exposes the site (Cleasby et al., 1992). Usually, substrates form an equilibrium between the monomeric, micellar, and emulsified states, which require a suitable model in order to study lipase kinet-ics. Sarda and Desnuelle (1958) documented the fact that interfacial activation is an enhancement of lipase activity and that lipase works effectively on the two-dimensional surface of the micelle. They proposed that lipase becomes activated

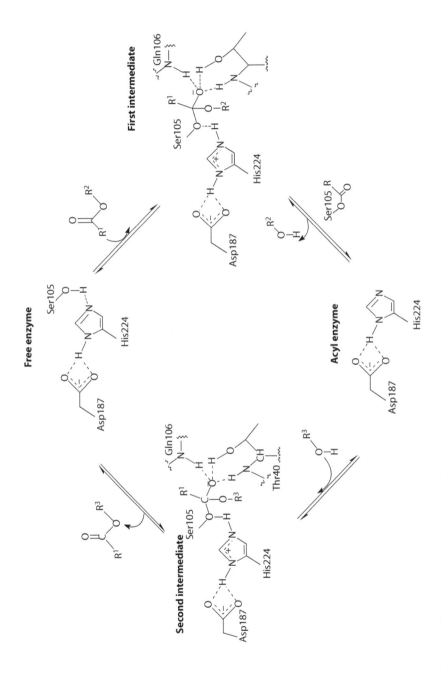

FIGURE 2.5 Lipase-catalyzed reaction mechanism.

when adsorbed at the water–lipid interface and is associated with a conformational change in the 3D structure (Desnuelle et al., 1960). During activation, the lid covering the active site moves and opens the binding sites to make the active sites accessible to the substrate. This provides a hydrophobic surface for the interaction. This lid movement forces the closed form to become open form and forms an oxyanion hole that increases lipase activity (Kumar, 2006). Open configuration and conformational changes were also found to be important in the interfacial activation of *P. cepacia* and *P. glumae* lipases (Schrag et al., 1997).

This interfacial activation does not appear in all hydrolytic enzymes. For example, cutinase from *fusarium solani*, which hydrolyzes cutin, usually has the fatty acids n-C_{16} and n-C_{18} and does not have a lid and does not exhibit these characteristics (Carvalho et al., 1998; Schrag et al., 1997). In addition, the presence of a lid does not always cause the enzyme to be active at the interface. The exception is lipase B from *Candida antarctica* that does not show interfacial activation despite having a lid (Kuo and Gardner, 2002; Uppenberg et al., 1994). Additionally, *Staphylococcus hyicus* and *P. aeruginosa* show an interfacial activation with some substrates but not with others (Verger, 1997).

X-ray studies of the 3D structure of lipase confirmed the existence of only one domain, except in pancreatic lipases where there were two distinct domains: a large N-terminal domain (amino-terminal domain) and a smaller C-terminal domain. The large N-terminal domain is a typical hydrolase that contains an active site with a catalytic triad formed by Ser, Asp, and His (van Tilbeurgh et al., 1992).

Since lipases catalyze long-chain triacylglycerolds, which are insoluble in water and form aggregates, their kinetics cannot be easily described by the Michaelis-Menten equation (discussed in Chapter 4). The large concentration of substrate needed to reach the critical micelle concentration (CMC) makes it difficult to easily define the reaction mixture. Due to the insolubility of the substrate, only those present at the interface and available to the lipase need to be considered. Thus, lipase activity is controlled by the micelle substrate at the interface and is independent of the molar concentration. This dependence on the interface surface was verified by the measuring the rate of hydrolysis of triglyceride in the presence of pancreatic lipase (Benzonana and Desnuelle, 1965). It was reported that lipase-catalyzed reactions can follow the Michaelis-Menten equation if the substrate concentration is expressed in terms of interfacial area per unit volume. The smaller the emulsion particles, the greater the total surface area for a given triglyceride quantity, therefore, the greater the reaction rate. In addition, it was found that lipase adsorbs reversibly to the emulsified particles and follows a Langmuir isotherm (Benzonana and Desnuelle, 1965). Once the adsorption takes place, activation follows, and the lipase takes on an active form. After that, the substrate fits to the active site and converts to the product. A similar model was used to describe the interfacial hydrolysis of short- and medium-chain lipids that produce water-soluble products (Verger, 1997; Verger and de Haas 1976; Verger et al., 1973). The first step is lipase fixation to the interface through an adsorption–desorption mechanism where lipase is initiated (E*) and binds to a substrate (S) forming a lipase-substrate complex (E*S), which then generates a product that is soluble in water, as shown in Equation 2.1:

$$E^* \underset{k_{-1}}{\overset{k_1}{\rightleftharpoons}} E^*S \xrightarrow{k_{cat}} P^*$$

with S entering at the top toward k_1, and $k_d \| k_p$ connecting E^* to E below.

$$\tag{2.1}$$

However, lipases also act on long-chain substrates that generate water-insoluble products, where the accumulation of those products at the interface might cause either surface substrate dilution, inhibition, or activation phenomena. Thus, a suitable model should consider the interfacial activation and generated insoluble product segregation. The most common models that have been proposed are the surface dilution model (Deems et al., 1975), describing lipolytic enzyme action on mixed micelles, and the scooting and hopping model. Equations 2.2 and 2.3 show the proposed interface model and the kinetics, respectively. These are similar to those that will be discussed in Chapter 4.

$$v = \frac{v_{max}[A][B]}{K_s^A K_M^B + K_M^B[A] + [A][B]} \tag{2.2}$$

$$E + A \rightleftharpoons EA + B \rightleftharpoons EAB \rightleftharpoons EA + P \tag{2.3}$$

where

$$K_s^A = \frac{k_{-1}}{k_1} \tag{2.4}$$

$$K_M^B = \frac{k_{-2} + k_3}{k_2} \tag{2.5}$$

where [A] is the molar concentration of the micellar sites that are capable of association with an enzyme, and [B] is the concentration of the substrate in the micelle surface in moles per unit area.

2.4   SPECIFICITY AS AN ATTRACTIVE FEATURE OF LIPASE

Lipases have several attractive properties that make them suitable for several applications. Lipase specificity eliminates the production of undesired products in the waste stream, decreases side reactions, and simplifies separation problems (Hasan et al., 2006; Pandey et al., 1999). Moreover, lipase reactions can be carried out under mild pH and temperature conditions, which reduces energy consumption and protects unstable substrates and products from denaturation. Another potential feature of lipases is their stability in an organic solvent, where they catalyze reactions without any aid from cofactors (Jaeger and Reetz, 1998). By choosing the proper lipase, the desired product generation can be controlled and unwanted side reactions are minimized.

2.4.1 POSITIONAL SPECIFICITY

Nonspecific lipases break down acylglycerol molecules in random positions, producing free fatty acids and glycerol with monoacylglycerols and diacylglycerols as intermediates from all three positions for triacylglycerol. Lipases from *Candida cylinderaceae* (Bello et al., 1987), *C. rugose* (Hermansyah et al., 2007; Pereira et al., 2001), *G. candidum, Staphylococcus aureus, Humicola lanuginose* (Sztajer and Zboińska, 1988), and *P. cepacia* (Chandler, 2001) are examples of such enzymes. On the other hand, 1,3-specific lipases breakdown fatty acids at positions *sn*-1 and *sn*-3 for acylglycerol and are unable to act on position *sn*-2 due to the steric hindrance that prevents position *sn*-2 fatty acid from entering the active site. *T. lanuginosus* (Rønne et al., 2005), *M. miehei* (Schuch and Mukherjee, 1987), and hog pancreas (Arcos et al., 2000) are examples of 1,3-specific lipases. Typically, pancreatic and fungal lipases are 1,3-positional-specific (Adlercreutz, 1994), whereas bacterial and yeast are nonpositional specific (Ghosh et al., 1996). However, *Bacillus stearothermophilus, Bacillus licheniformis*, and *Bacillus subtilus*, which are bacterial lipases, showed 1,3-positional specificity (Chen et al., 2004). *Sn*-2 specific lipases release fatty acid only from position 2 of the acylglycerol backbone. Commonly, positional specificity is determined by partial hydrolysis of olive oils and cocoa butter (which contains mainly palmatic, oleic, and steric acids) and the separation by thin-layer chromatography for subsequent analysis (Van Camp et al., 1998).

2.4.2 SUBSTRATE SPECIFICITY

Lipases are divided into three groups based on their binding sites. The groups are (1) lipases with crevice-like binding sites such as lipases from *Rhizomucor* and *Rhizopus*; (2) lipases with a tunnel-like binding site such as lipases from *C. rugosa*; and (3) lipases with funnel-like binding sites such as lipases from *C. antarctica* and *P. cepacia* (Pleiss et al., 1998), where the shape and the microenvironment of the binding site determine lipase specificity.

2.4.3 FATTY ACID SPECIFICITY

Many lipases are specific to certain fatty acid substrates, with some being active toward specific long-chain fatty acids, whereas others are specific to medium- and short-chain fatty acids. Thus, lipases are able to distinguish between structures. *P. aeruginosa* EF2 (Gilbert, Cornish et al., 1991; Gilbert, Drozd et al., 1991) and *P. aeruginosa* PAC1R (Misset et al., 1994) show specificity toward long-chain fatty acids, whereas lipases from *Bacillus subtilis* 168 (Lesuisse et al., 1993) and *Bacillus* sp. THL027 (Dharmsthiti and Lchai, 1999) favor small- or medium-chain fatty acids. Lipase from *Staphylococcus aureus* 226 displays a preference for unsaturated fatty acids (Muraoka et al., 1982).

2.4.4 ENANTIO-STEREOSPECIFICITY

Stereospecificity is another important feature of lipases. Some lipases, in addition to being able to distinguish between *sn*-1 and *sn*-2 triglycerides, can also distinguish between enantioisomers from chiral molecules. This feature has applications in the biotechnology and pharmaceutical industries in producing pure chiral isomers for drug syntheses (Patel, 2000; Zhao and Chen, 2008).

2.5 LIPASES PRODUCTION, ISOLATION, AND PURIFICATION

Lipases are produced from different sources. Most of the commercially produced lipases come from microbial origins. Several methods have been suggested for screening in lipase production. This involves the plate and broth assay (Sierra, 1957). In the plate detection assay, agar plates that contain Tween 80 as substrate in solid medium are used. In such a method, cloudy zone formation around the colonies was an indication of lipase production. Tributrine, Trioline and Rhodamineb are other substrates where clear zones around the colonies indicate lipase production on agar plates (Cardenas et al., 2001; Wang et al., 1995). In the broth assays, lipase activity is assayed according to the colorimetric method (Kwon and Rhee, 1986).

Lipases are mainly produced by both submerged and solid state fermentation and most of them are of the extracellular type, which facilitates the purification processes. Many studies have been carried out to identify the optimal culture and requirements for lipase production, where the production is strongly influenced by many nutritional and environmental parameters such as temperature, pH, nitrogen and carbon sources, aeration, and dissolved oxygen (Espinosa et al., 1990; Nashif and Nelson, 1953). Enzyme production by means of solid-state fermentation has the advantage of reducing production costs. This has been proven by a comprehensive economic analysis study (de Azeredo et al., 2007), which showed that lipase produced from *Penicillium restrictum* by solid-state fermentation is more economically feasible than production by submerged fermentation. It was shown that a 68% reduction in the cost of the solid state can be achieved using fermentation.

Fermentation is usually followed by the removal of cells from the culture broth, either by centrifugation or by filtration (Westoby et al., 2011). The cell-free broth is then concentrated by ultrafiltration, ammonium sulfate precipitation, or extraction with an organic solvent (Saxena et al., 2003). A single purification step is usually not adequate to get lipase of high purity. Therefore, purification using a combination of chromatographic techniques is mainly employed. However, except in the pharmaceutical industry, pure lipases are required and crude enzymes are usually adequate.

For enzyme purification, ion exchange, gel filtration, and affinity chromatography are the most commonly reported techniques, with ion exchange being the most commonly employed method. Diethylaminoethyl groups in anion exchange and carboxymethyl in cation exchange are the most commonly used ion exchangers (Hagel et al., 2008). Strictly speaking, affinity chromatography can be used prior to ion exchange and gel filtration, however, the materials are expensive. These purification methods are time consuming and may reduce the final yield (Saxena et al., 2003). Thus, novel techniques have been applied, such as aqueous two-phase, cross-flow membrane filtration and immunopurification, which leads to purification of 1000- to 10,000-fold in a single step (Belguith et al., 2009).

2.6 INDUSTRIAL APPLICATIONS OF LIPASE

Currently, lipases are used in a great number of fields due to their unique properties such as enantioselectivity, region selectivity, and specificity. They are one of the most important biocatalysts for biotechnological applications in catalyzing different reactions such as esterification, interesterfication, acidolysis, transesterification, and aminolysis. A comparison between these reactions is shown in Figure 2.6. This makes them the preferred catalysts for a number of industrial applications, such as in detergents, food, pharmaceuticals, leather, textiles, cosmetics, and paper, with the most significant applications being the first three (Houde et al., 2004). The main restriction on lipase use is the high costs associated with their production. This can be overcome by immobilization and molecular technology. Further details of immobilized lipases are presented in Chapter 3.

2.6.1 DETERGENTS

The major commercial application of lipases is in detergents for household uses, laundry, and dishwashers. Lipases are able to hydrolyze fats of various compositions under severe washing conditions to monoglycerides, diglycerides, glycerol, and free fatty acids, which are more soluble than the original fats. The high level of activity of lipase at mild temperatures reduces energy consumption, maintains the quality of clothing fabrics, and reduces any harmful residues. To be used in detergents, lipases need to be thermostable and remain active in the alkaline environment. Lipases from *Acinetobacter radioresistens* and *Bacillus* sp. showed a stability of pH of 10 (Chen et al. 1998; Hasan et al., 2006).

(a) $R_1COOR_2 + H_2O \rightleftharpoons R_1COOH + R_2OH$

(b) $R_1COOH + R_2OH \rightleftharpoons R_1COOR_2 + H_2O$

(c) $R_1COOR_2 + R_3COOR_4 \rightleftharpoons R_3COOR_2 + R_1COOR_4$

(d) $R_1COOR_2 + R_3COOH \rightleftharpoons R_1COOR_3 + R_2COOH$

(e) $R_1COOR_2 + R_3OH \rightleftharpoons R_1COOR_3 + R_2OH$

(f) $R_1COOR_2 + R_3NH_2 \rightleftharpoons R_1COONHR_3 + R_2OH$

FIGURE 2.6 Reactions catalyzed by lipases (a) hydolysis, (b) esterification, (c) interesterification, (d) acidolysis, (e) alcoholysis, and (f) aminolysis.

Novo Nordisk, most recently known as Novozymes, developed a lipase in 1988 produced from *H. lanuginose* to be used in detergents. However, the production quantity was too low for commercialization. In 1994, Lipolase® was developed as the first commercial fungus recombinant lipase, which was isolated from *H. lanuginose* and was expressed in *Aspergillus oryzae* (Boel et al., 1994). In the following year, Lumafast from *P. mendocina* and Lipomax from *Pseudomonas alcaligenes* bacterial lipases were produced by Genencor International (Fariha et al., 2010; Jaeger and Reetz, 1998), and were also used in detergents.

2.6.2 LEATHER

During the processing of hides and skins, the removal of residual fats is essential. The removal is carried out by a chemical process called liming, which is far from efficient. In addition, solvents and surfactants are used in the degreasing process to get the required quality of product. These chemicals produce large amounts of waste, are which have a negative impact on the environment. Recently, lipases have been employed as an alternative for fat and grease removal from leather. For example, *R. nodosus* lipase has been successfully used for the degreasing of suede clothing and sheepskins (Hasan et al., 2006).

2.6.3 FOOD PROCESSING

Fats and oils are important food constituents. Their nutritional, chemical, and physical properties are influenced by the position of fatty acids, their chain length, and the degree of unsaturation. Usually, lipases are used to obtain modified fats with nutritionally improved properties and they provide high value fats such as cocoa butter that contains palmitic and stearic acids. Commercial lipases are mainly employed in the dairy industry for flavor enhancement in cheese (Mase et al., 2010; Omar et al., 1986), the acceleration of cheese ripening (Fox et al., 1996; Kheadr et al., 2002), the manufacture of cheeselike products, and the lipolysis of butterfat and cream (Purko et al., 1952; Seitz, 1974). Lipases release short-chain fatty acids that develop a tangy flavor and medium chain fatty acids that give a soapy taste to the end product (Sharma et al., 2011).

Lipases are also employed in processing tea. Processing fresh leaves into black tea depends on the hydrolytic enzymes in the leaf. Several biochemical changes take place before being converted to black tea. Lipases are one of the enzymes that are involved in this process by liberating free fatty acids in the leaves. This later initiates the formation of volatile compounds using lipoxygenase (Ramarethinam et al., 2002). *R. miehei* lipase has been tested for its ability in enhancing the aroma during the processing of tea (Ramarethinam et al., 2002).

2.6.4 PULP AND PAPER

Wood is the main raw material used in the production of paper. In the pulping process, wood extractives such as triglycerides and fatty acids, which reduce paper quality and machine efficiency, are produced. Pitch sticks on the metal parts of paper-making machines, including their rolls and wires.

Traditionally, pitch problems were chemically controlled by the adsorption and dispersion of pitch particles by adding fine talc, dispersants, and other chemicals. However, this use of chemicals leads to problems with effluent treatment and environmental pollution. Seasoning of the wood before pulping is another method where the wood is left outdoors for several months. However, these affect the pulp's brightness and yield, which results in an increase in capital cost and land use. Enzymatic control has proven to be a successful alternative, where lipases have shown their ability to reduce the triglyceride content (Farrell et al., 1997). Lipase AYL, produced from *C. rugosa* and from *C. cylindracea*, has been successfully used in the pulp and paper industry (Singh and Mukhopadhyay, 2011).

2.6.5 WASTEWATER TREATMENT

Fats present in wastewater are nondegradable, causing the generation of unpleasant odors, foam formation, and solidification at low temperatures (Rigo et al., 2008). Lipases have potential applications in wastewater treatment, as they reduce fat and minimize by-product formation by using mild conditions, and reducing energy needs and costs.

2.6.6 BIODIESEL PRODUCTION

Lipase's role in organic chemicals synthesis has received great attention due to its numerous advantages over chemical catalysts. It has been used in many chemical syntheses including biodiesel production. Biodiesel is a monoalkyl ester made of vegetable oils and animal fats via transesterification reactions with alcohol, normally methanol. An alkali catalysis process has been established that records a high conversion rate for oils, actually reaching 98%. Alkali catalysts such as sodium hydroxide (NaOH), potassium hydroxide (KOH), and sodium methoxide (CH_3ONa) have been used successfully. However, the most important limitation is the process's sensitivity to both free fatty acids and water content. Acid catalysis transesterification has been proposed to solve free fatty acid limitation, but it is not as popular as the alkali-catalyzed process due to corrosion and a slow reaction rate. To overcome conventional chemical catalysts limitations, lipases have been proposed (Al-Zuhair et al., 2007; Fukuda et al., 2001). The most important advantage of using lipases is their ability to completely convert free fatty acids contained in the fat or oil to methyl esters. Additionally, glycerol and any by-product can be easily recovered (Demirbas, 2009; Li et al., 2006; Modi et al., 2007).

2.7 IMPROVEMENTS OF LIPASE PROPERTIES

Despite the various uses for lipases, the availability of lipases with specific characteristics, their stability, and operational properties limits their uses in many applications, especially in the biosynthesis of molecules in organic media. Researchers have managed to overcome some of these shortcomings through reaction medium engineering and lipase engineering physically, chemically, or genetically.

2.7.1 MEDIUM ENGINEERING

The reaction medium can be improved by modifying the medium polarity. This determines the solubility and stability of lipase substrates, and therefore, lipase activity and selectivity. Introducing organic solvents to synthetic reactions increases the solubility of poor water-soluble substrates and so increases reaction rates. The rates of lipase-catalyzed reactions are strongly dependent on the water activity of the reaction medium (Berglund, 2001), and reducing water activity diminishes hydrolysis during transesterification reactions (Radzi et al., 2005). Nevertheless, organic solvents may denature or inhibit the lipase and increase reaction complexity. These aspects also have to be considered when choosing a suitable solvent, including solvent compatibility, inertness, low density, toxicity, and flammability (Marek and Sajja, 2004). In this regard, lipases have been tested in organic solvents such as *tert*-butanol, *n*-hexane (Eltaweel et al., 2005; Hernández-Rodríguez et al., 2009), ionic liquids (Kaar et al., 2003; Ulbert et al., 2005), and supercritical fluids (Knez and Habulin, 2002; Nakamura et al., 1986; Palocci et al., 2008).

2.7.2 LIPASE ENGINEERING

Lipases enantioselectivity can be improved physically, chemically, or genetically. Physically, lipases can be modified by cross-linking, crystallization, and immobilization that involves attaching the lipase onto an insoluble solid support. Immobilization enhances the stability of lipase and facilitates lipase recovery. Details of lipase immobilization methods are discussed in Chapter 3.

The chemical modification of lipase can be carried out using modifiers that interact with certain amino acids. Modifiers can modify the amino acid residues and covalently bind the lipase to water-insoluble materials and matrices. Porcine pancreatic lipase catalytic activity was enhanced by reductive alkylation of the amino carried by a reaction of the amino group with an aldehyde or ketone that resulted in about a 50% increase in the v_{max} value of the lipase (Kaimal and Saroja, 1989). Reactions between functional groups of the protein and reactive groups on the activated material covalently bind the lipase with water-insoluble materials and enhance lipase properties. Detergent-modified lipase from *R. delemar* synthesized using didodecylglucosyl glutamate synthetic detergent is an example of modified lipases by covalent coupling that show an improvement in lipase thermostability and reactivity (Tsuzuki and Suzuki, 1991).

Genetic engineering of lipase involves modifying the gene encoding the lipase. Through genetic engineering, nucleic acids can be isolated and manipulated, amino acid sequences can be altered by allowing the insertion or deletion of certain amino acids, and segments from different genes and different organisms can be combined. Properties can be altered by site-directed mutagenesis, directed evolution, or metagenome approaches. This is complicated and often a matter of trial and error, plus a detailed understanding of substrate–lipase interactions is required. Lipase specificity for short- or medium-chain fatty acids of *R. delemar* was improved through site-directed mutagenesis (Klein et al., 1997). An improvement in substrate specificity was achieved from *P. mendocina* lipases through mutagenic alteration of amino acids in the immediate vicinity of the catalytic triad, where single mutations generate a group of mutant lipases whose k_{cat}/K_m values against short- and medium-chain-length esters varied over a hundredfold (Bott et al., 1994).

Directed evolution is a powerful tool used to improve lipase properties that does not depend on a comprehensive understanding of the relationship between enzyme structure and function. It rather depends on simple, yet powerful, random mutation and selection. The targeted genes are exposed to iterative cycles of random mutagenesis, expressed in an appropriate host and subsequently screened (Johannes and Zhao, 2006). *Bacillus* lipase was engineered by directed evolution, where a lip gene was cloned and expressed in *E. coli*. The mutagenesis was executed by error-prone polymerase chain reaction (PCR). The mutation enhanced the specific activity of the lipase by twofold (Khurana et al., 2011).

REFERENCES

Adlercreutz, P. 1994. "Enzyme-Catalysed Lipid Modification." *Biotechnology & Genetic Engineering Reviews* 12:231–254.

Al-Zuhair, S., F. W. Ling, and L. S. Jun. 2007. "Proposed Kinetic Mechanism of the Production of Biodiesel from Palm Oil Using Lipase." *Process Biochemistry* 42 (6):951–960.

Arcos, J., H. Garcia, and C. Hill. 2000. "Regioselective Analysis of the Fatty Acid Composition of Triacylglycerols with Conventional High-Performance Liquid Chromatography." *Journal of the American Oil Chemists' Society* 77 (5):507–512.

Belguith, H., S. Fattouch, T. Jridi, and J. Ben Hamida. 2009. "Immunopurification of a Rape (*Brassica Napus* L.) Seedling Lipase." *African Journal of Biochemistry Research* 3:356–365.

Bello, M., D. Thomas, and M. D. Legoy. 1987. "Interesterification and Synthesis by *Candida Cylindracea* Lipase in Microemulsions." *Biochemical and Biophysical Research Communications* 146 (1):361–367.

Benzonana, G., and P. Desnuelle. 1965. "Kinetic Study of the Action of Pancreatic Lipase on Emulsified Triglycerides. Enzymology Assay in Heterogeneous Medium." *Biochimica et Biophysica Acta* 105 (1):121–136.

Berglund, P. 2001. "Controlling Lipase Enantioselectivity for Organic Synthesis." *Biomolecular Engineering* 18 (1):13–22.

Boel, E., T. Christensen, and H. Wöldike. 1994. "Process for the Production of Protein Products in *Aspergillus*." Novozymes A/S, Bagsvaerd, Denmark.

Bott, R., J. W. Shield, and A. J. Poulose. 1994. "Protein Engineering of Lipases." In *Lipases: Their Structure, Biochemistry and Application*, edited by P. Woolley and S. B. Petersen. Cambridge University Press.

Brady, L., A. M. Brzozowski, Z. S. Derewenda, E. Dodson, S. Tolley, J. P. Turkenburg, L. Christiansen et al. 1990. "A Serine Protease Triad Forms the Catalytic Centre of a Triacylglycerol Lipase." *Nature* 343 (6260):767–770.

Cardenas, F., E. Alvarez, M. S. de Castro-Alvarez, J. M. Sanchez-Montero, M. Valmaseda, S. W. Elson, and J. V. Sinisterra. 2001. "Screening and Catalytic Activity in Organic Synthesis of Novel Fungal and Yeast Lipases." *Journal of Molecular Catalysis B: Enzymatic* 14 (4):111–123.

Carvalho, C. M. L. C., M. R. Aires-Barro, and J. M. S. Cabral. 1998. "Cutinase Structure, Function and Biocatalytic Applications." *Electronic Journal of Biotechnology* 1:160–173.

Chahinian, H., and L. Sarda. 2009. "Distinction between Esterases and Lipases: Comparative Biochemical Properties of Sequence-Related Carboxylesterases." *Protein & Peptide Letters* 16 (10):1149–1161.

Chandler, I. 2001. "Determining the Regioselectivity of Immobilized Lipases in Triacylglycerol Acidolysis Reactions." *Journal of the American Oil Chemists' Society* 78 (7):737–742.

Chen, S., C. Cheng, and T. Chen. 1998. "Production of an Alkaline Lipase by *Acinetobacter Radioresistens.*" *Journal of Fermentation and Bioengineering* 86 (3):308–312.

Chen, L., T. Coolbear, and R. M. Daniel. 2004. "Characteristics of Proteinases and Lipases Produced by Seven *Bacillus* Sp. Isolated from Milk Powder Production Lines." *International Dairy Journal* 14 (6):495–504.

Cheng, M., C. Angkawidjaja, Y. Koga, and S. Kanaya. 2014. "Calcium-Independent Opening of Lid1 of a Family I.3 Lipase by a Single Asp to Arg Mutation at the Calcium-Binding Site." *Protein Engineering Design and Selection* 27 (5):169–176.

Cleasby, A., E. Garman, M. R. Egmond, and M. Batenburg. 1992. "Crystallization and Preliminary X-Ray Study of a Lipase from *Pseudomonas Glumae.*" *Journal of Molecular Biology* 224:281–282.

Comménil, P., L. Belingheri, M. Sancholle, and B. Dehorter. 1995. "Purification and Properties of an Extracellular Lipase from the Fungus *Botrytis Cinerea.*" *Lipids* 30 (4):351–356.

Cygler, M., and J. D. Schrag. 1999. "Structure and Conformational Flexibility of *Candida Rugosa* Lipase." *Biochimica et Biophysica Acta (BBA)—Molecular and Cell Biology of Lipids* 1441 (2–3):205–214.

de Azeredo, L. A. I., P. M. Gomes, G. L. Sant'Anna, L. R. Castilho, and De. M. G. Freire. 2007. "Production and Regulation of Lipase Activity from *Penicillium Restrictum* in Submerged and Solid-State Fermentations." *Current Microbiology* 54 (5):361–365.

Deems, R. A., B. R. Eaton, and E. A. Dennis. 1975. "Kinetic Analysis of Phospholipase A2 Activity toward Mixed Micelles and Its Implications for the Study of Lipolytic Enzymes." *Journal of Biological Chemistry* 250 (23):9013–9020.

Demirbas, A. 2009. "Progress and Recent Trends in Biodiesel Fuels." *Energy Conversion and Management* 50 (1):14–34.

Desnuelle, P., L. Sarda, and G. Ailhard. 1960. "Inhibition De La Lipase Pancréatique Par Le Diéthyl-P-Nitrophényl Phosphate En Emulsion." *Biochimica et Biophysica Acta* 37:570–571.

Dharmsthiti, S., and S. Lchai. 1999. "Production, Purification and Characterization of Thermophilic Lipase from *Bacillus* Sp. Thl027." *FEMS Microbiology Letters* 179 (2):241–146.

Eltaweel, M. A., R. N. Z. R. Abd Rahman, A. B. Salleh, and M. Basri. 2005. "An Organic Solvent-Stable Lipase from *Bacillus* Sp. Strain 42." *Annals of Microbiology* 55 (3):187–192.

Espinosa, E., S. Sánchez, and A. Farrés. 1990. "Nutritional Factors Affecting Lipase Production by *Rhizopusdelemar* CDBB H313." *Biotechnology Letters* 12 (3):209–214.

Fariha, H., A. S. Aamer, J. Sundus, and H. Abdul. 2010. "Enzymes Used in Detergents: Lipases." *African Journal of Biotechnology* 9 (31):4836–4844.

Farrell, R. L., K. Hata, and M. B. Wall. 1997. "Solving Pitch Problems in Pulp and Paper Processes by the Use of Enzymes or Fungi." *Advances in Biochemical Engineering/ Biotechnology & Genetic Engineering Reviews* 57:197–211.

Fojan, P., P. H. Jonson, M. T. N. Petersen, and S. B. Petersen. 2000. "What Distinguishes an Esterase from a Lipase: A Novel Structural Approach." *Biochimie* 82 (11):1033–1041.

Fox, P. F., J. M. Wallace, S. Morgan, C. M. Lynch, N. J. Niland, and J. Tobin. 1996. "Acceleration of Cheese Ripening." *Antonie van Leeuwenhoek* 70 (2):271–297.

Fukuda, H., A. Kondo, and H. Noda. 2001. "Biodiesel Fuel Production by Transesterification of Oils." *Journal of Bioscience and Bioengineering* 92 (5):405–416.

Ghosh, P. K., R. K. Saxena, R. Gupta, R. P. Yadav, and S. Davidson. 1996. "Microbial Lipases: Production and Applications." *Science Progress* 79:119–157.

Ghosh, M., S. Bhattacharyya, and D. K. Bhattacharyya. 2005. "Production of Lipase and Phospholipase Enzymes from *Pseudomonas* Sp. And Their Action on Phospholipids." *Journal of Oleo Science* 54 (7):407–411.

Gilbert, E. J., A. Cornish, and C. W. Jones. 1991. "Purification and Properties of Extracellular Lipase from *Pseudomonas Aeruginosa* EF2." *Journal of General Microbiology* 137 (9):2223–2229.

Gilbert, E. J., J. W. Drozd, and C. W. Jones. 1991. "Physiological Regulation and Optimization of Lipase Activity in *Pseudomonas Aeruginosa* EF2." *Journal of General Microbiology* 137 (9):2215–2221.

Grochulski, P., Y. Li, J. D. Schrag, F. Bouthillier, P. Smith, D. Harrison, B. Rubin, and M. Cygler. 1993. "Insights into Interfacial Activation from an Open Structure of *Candida-Rugosa* Lipase." *Journal of Biological Chemistry* 268:12843–12847.

Gulomova, K., E. Ziomek, J. D. Schrag, K. Davranov, and M. Cygler. 1996. "Purification and Characterization of *a Penicillium* Sp. Lipase Which Discriminates against Diglycerides." *Lipids* 31 (4):379–384.

Haas, M., D. Cichowicz, and D. Bailey. 1992. "Purification and Characterization of an Extracellular Lipase from the Fungus *Rhizopus Delemar.*" *Lipids* 27 (8):571–576.

Hagel, L., G. Jagschies, and G. K. Sofer. 2008. *Handbook of Process Chromatography: Development, Manufacturing, Validation and Economics.* Academic Press.

Hasan, F., A. A. Shah, and A. Hameed. 2006. "Industrial Applications of Microbial Lipases." *Enzyme and Microbial Technology* 39 (2):235–251.

Hermansyah, H., A. Wijanarko, Dianursanti, M. Gozan, P. P. D. K. Wulan, R. Arbianti, R. W. Soemantojo et al. 2007. "Kinetics Model for Triglyceride Hydrolysis Using Lipase: Review." *Makara Journal of Technology* 11 (1):30–35.

Hernández-Rodríguez, B., J. Córdova, E. Bárzana, and E. Favela-Torres. 2009. "Effects of Organic Solvents on Activity and Stability of Lipases Produced by Thermotolerant Fungi in Solid-State Fermentation." *Journal of Molecular Catalysis B: Enzymatic* 61 (3–4):136–142.

Holmquist, M. 2000. "Alpha Beta-Hydrolase Fold Enzymes Structures, Functions and Mechanisms." *Current Protein & Peptide Science* 1 (2):209–235.

Houde, A., A. Kademi, and D. Leblanc. 2004. "Lipases and Their Industrial Applications." *Applied Biochemistry and Biotechnology* 118 (1):155–170.

Jacobsen, T., J. Olsen, and K. Allermann. 1990. "Substrate Specificity of *Geotrichum Candidum* Lipase Preparations." *Biotechnology Letters* 12 (2):121–126.

Jaeger, K. E., and M. T. Reetz. 1998. "Microbial Lipases Form Versatile Tools for Biotechnology." *Trends in Biotechnology* 16 (9):396–403.

Jaeger, K. E., S. Ransac, B. W. Dijkstra, C. Colson, M. van Heuvel, and O. Misset. 1994. "Bacterial Lipases." *FEMS Microbiology Reviews* 15 (1):29–63.

Jaeger, K. E., B. W. Dijkstra, and M. T. Reetz. 1999. "Bacterial Biocatalysts: Molecular Biology, Three-Dimensional Structures, and Biotechnological Applications of Lipases." *Annual Review of Microbiology* 53:315–351.

Johannes, T. W., and H. Zhao. 2006. "Directed Evolution of Enzymes and Biosynthetic Pathways." *Current Opinion in Microbiology* 9 (3):261–267.

Kaar, J. L., A. M. Jesionowski, J. A. Berberich, R. Moulton, and A. J. Russell. 2003. "Impact of Ionic Liquid Physical Properties on Lipase Activity and Stability." *Journal of the American Chemical Society* 125 (14):4125–4131.

Kaimal, T. N. B., and M. Saroja. 1989. "Enhancement of Catalytic Activity of Porcine Pancreatic Lipase by Reductive Alkylation." *Biotechnology Letters* 11 (1):31–36.

Kheadr, E. E., J. C. Vuillemard, and S. A. El-Deeb. 2002. "Acceleration of Cheddar Cheese Lipolysis by Using Liposome-Entrapped Lipases." *Journal of Food Science* 67 (2):485–492.

Khurana, J., R. Singh, and J. Kaur. 2011. "Engineering of *Bacillus* Lipase by Directed Evolution for Enhanced Thermal Stability: Effect of Isoleucine to Threonine Mutation at Protein Surface." *Molecular Biology Reports* 38 (5):2919–2926.

Klein, R., G. King, R. Moreau, and M. Haas. 1997. "Altered Acyl Chain Length Specificity of *Rhizopus Delemar* Lipase through Mutagenesis and Molecular Modeling." *Lipids* 32 (2):123–130.

Knez, Ž., and M. Habulin. 2002. "Compressed Gases as Alternative Enzymatic-Reaction Solvents: A Short Review." *The Journal of Supercritical Fluids* 23 (1):29–42.

Kojima, Y., M. Yokoe, and T. Mase. 1994. "Purification and Characterization of an Alkaline Lipase from *Pseudomonas Fluorescens* Ak102." *Bioscience, Biotechnology, and Agrochemistry* 49 (9):1564–1568.

Kumar, A. 2006. *Protein Biotechnology*. Discovery Publishing House.

Kundu, M., J. Basu, and M. Guchhait. 1987. "Isolation and Characterization of an Extracellular Lipase from the Conidia of *Neurospora Crassa*." *Journal of General Microbiology* 133 (1):149–153.

Kuo, T. M., and H. W. Gardner. 2002. *Lipid Biotechnology*. Marcel Dekker.

Kwon, D., and J. Rhee. 1986. "A Simple and Rapid Colorimetric Method for Determination of Free Fatty Acids for Lipase Assay." *Journal of the American Oil Chemists' Society* 63 (1):89–92.

Lesuisse, E., K. Schanck, and C. Colson. 1993. "Purification and Preliminary Characterization of the Extracellular Lipase of *Bacillus Subtilis* 168, an Extremely Basic pH-Tolerant Enzyme." *European Journal of Biochemistry* 216 (1):155–160.

Li, H., and X. Zhang. 2005. "Characterization of Thermostable Lipase from Thermophilic *Geobacillus* Sp. Tw1." *Protein Expression and Purification* 42 (1):153–159.

Li, L., W. Du, D. Liu, L. Wang, and Z. Li. 2006. "Lipase-Catalyzed Transesterification of Rapeseed Oils for Biodiesel Production with a Novel Organic Solvent as the Reaction Medium." *Journal of Molecular Catalysis B: Enzymatic* 43 (1–4):58–62.

Lin, S., J. Lee, and C. Chiou. 1996. "Purification and Characterization of a Lipase from *Neurospora* Sp. Tt-241." *Journal of the American Oil Chemists' Society* 73 (6):739–745.

Mala, J. G. S., and S. Takeuchi. 2008. "Understanding Structural Features of Microbial Lipases: An Overview." *Journal of Analytical Chemistry Insights* 3:9–19.

Marek, A., and H. K. Sajja. 2004. "Strategies for Improving Enzymes for Efficient Biocatalysis." *Food Technology and Biotechnology* 42 (4):251–264.

Mase, T., M. Sonoda, M. Morita, and E. Hirose. 2010. "Characterization of a Lipase from *Sporidiobolus Pararoseus* 25-a Which Produces Cheese Flavor." *Food Science and Technology Research* 17 (1):17–20.

Meyer, J. G., J. G. Renczes, and H. Kaffarnik. 1990. "Lipolytic Enzymes of the Human Pancreas." *Journal of Molecular Medicine* 68 (2):60–64.

Miled, N., A. De Caro, J. De Caro, and R. Verger. 2000. "A Conformational Transition between an Open and Closed Form of Human Pancreatic Lipase Revealed by a Monoclonal Antibody." *Biochimica et Biophysica Acta (BBA)—Protein Structure and Molecular Enzymology* 1476 (2):165–172.

Misset, O., G. Gerritse, K. E. Jaeger, U. Winkler, C. Colson, K. Schanck, E. Lesuisse et al. 1994. "The Structure Function Relationship of the Lipases from *Pseudomonas Aeruginosa* and Bacillus Subtilis." *Protein Engineering* 7 (4):523–529.

Modi, M. K., J. R. C. Reddy, B. V. S. K. Rao, and R. B. N. Prasad. 2007. "Lipase-Mediated Conversion of Vegetable Oils into Biodiesel Using Ethyl Acetate as Acyl Acceptor." *Bioresource Technology* 98 (6):1260–1264.

Mukherjee, K. D. 1994. "Plant Lipases and Their Application in Lipid Biotransformations." *Progress in Lipid Research* 33 (1–2):165–174.

Muraoka, T., T. Ando, and H. Okuda. 1982. "Purification and Properties of a Novel Lipase from *Staphylococcus Aureus* 226." *Journal of Biochemistry* 92 (6):1933–1939.

Nakamura, K., Y. M. Chi, Y. Yamada, and T. Yano. 1986. "Lipase Activity and Stability in Supercritical Carbon Dioxide" *Chemical Engineering Communications* 45 (1–6):207–212.

Nardini, M., and B. W. Dijkstra. 1999. "α/β Hydrolase Fold Enzymes: The Family Keeps Growing." *Current Opinion in Structural Biology* 9 (6):732–737.

Nashif, S. A., and F. E. Nelson. 1953. "The Lipase of *Pseudomonas Fragil* II. Factors Affecting Lipase Production." *Journal of Dairy Science* 36 (5):471–480.

Noble, M. E. M., A. Cleasby, L. N. Johnson, M. R. Egmond, and L. G. J. Frenken. 1993. "The Crystal Structure of Triacylglycerol Lipase from *Pseudomonas Glumae* Reveals a Partially Redundant Catalytic Aspartate." *FEBS Letters* 331 (1–2):123–128.

Ollis, D. L., E. Cheah, M. Cygler, B. Dijkstra, F. Frolow, S. M. Franken, M. Harel et al. 1992. "The α/β Hydrolase Fold." *Protein Engineering Design and Selection* 5 (3):197–211.

Omar, M. M., A. I. El-Zayat, and M. Ashour. 1986. "Flavor Enhancement, by Lipase Addition, of Ras Cheese Made from Reconstituted Milk." *Food Chemistry* 19 (4):277–286.

Palocci, C., M. Falconi, L. Chronopoulou, and E. Cernia. 2008. "Lipase-Catalyzed Regioselective Acylation of Tritylglycosides in Supercritical Carbon Dioxide." *Journal of Supercritical Fluids* 45 (1):88–93.

Pandey, A., S. Benjamin, C. R. Soccol, P. Nigam, N. Krieger, and V. T. Soccol. 1999. "The Realm of Microbial Lipases in Biotechnology." *Biotechnology and Applied Biochemistry* 2:119–131.

Patel, R. N. 2000. "Microbial/Enzymatic Synthesis of Chiral Drug Intermediates." *Advances in Applied Microbiology* 47:33–78.

Pereira, E., H. De Castro, F. De Moraes, and G. Zanin. 2001. "Kinetic Studies of Lipase from *Candida Rugosa*." *Applied Biochemistry and Biotechnology* 91–93 (1):739–752.

Petersen, S. B., and F. Drabløs. 1994. "A Sequence Analysis of Lipases, Esterases and Related Proteins." In *Lipases: Their Structure, Biochemistry and Application*, edited by P. Woolley and S. B. Petersen. Cambridge University Press.

Pleiss, J., M. Fischer, and R. D. Schmid. 1998. "Anatomy of Lipase Binding Sites: The Scissile Fatty Acid Binding Site." *Chemistry and Physics of Lipids* 93 (1–2):67–80.

Pratuangdejkul, J., and S. Dharmsthiti. 2000. "Purification and Characterization of Lipase from Psychrophilic *Acinetobacter Calcoaceticus* Lp009." *Microbiological Research* 155 (2):95–100.

Purko, M., W. O. Nelson, and W. A. Wood. 1952. "The Liberation of Water-Insoluble Acids in Cream by *Geotrichum Candidum*." *Journal of Dairy Science* 35 (4):298–304.

Radzi, S., M. Basri, A. Salleh, A. Arif, R. Mohammad, M. B. Abdul Rahman, and R. N. Z. R. Abdul Rahman. 2005. "High Performance Enzymatic Synthesis of Oleyl Oleate Using Immobilised Lipase from Candida Antarctica." *Electronic Journal of Biotechnology* 8 (3):291–298.

Ramarethinam, S., K. Latha, and N. Rajalakshmi. 2002. "Use of a Fungal Lipase for Enhancement of Aroma in Black Tea." *Food Science and Technology Research* 8 (4):328–332.

Reetz, M. T. 2002. "Lipases as Practical Biocatalysts." *Current Opinion in Chemical Biology* 6 (2):145–150.

Reis, P., K. Holmberg, H. Watzke, M. E. Leser, and R. Miller. 2009. "Lipases at Interfaces: A Review." *Advances in Colloid and Interface Science* 147–148:237–250.

Rigo, E., R. E. Rigoni, P. Lodea, D. De Oliveira, D. M. G. Freire, H. Treichel, and M. Di Luccio. 2008. "Comparison of Two Lipases in the Hydrolysis of Oil and Grease in Wastewater of the Swine Meat Industry." *Industrial & Engineering Chemistry Research* 47 (6):1760–1765.

Rønne, T., L. Pedersen, and X. Xu. 2005. "Triglyceride Selectivity of Immobilized *Thermomyces Lanuginosa* Lipase in Interesterification." *Journal of the American Oil Chemists' Society* 82 (10):737–743.

Salameh, M. A., and J. Wiegel. 2010. "Effects of Detergents on Activity, Thermostability and Aggregation of Two Alkalithermophilic Lipases from *Thermosyntropha Lipolytica*." *Open Biochemistry Journal* 4:22–28.

Salleh, A., N. Z. R. Abdul Rahman, and M. Basri. 2006. "Lipases: Introduction." In *New Lipases and Proteases*. Nova Science.

Sarda, L., and P. Desnuelle. 1958. "Actions of Pancreatic Lipase on Esters in Emulsions." *Biochemica et Biophysica Acta* 30 (3):513–521.

Saxena, R. K., A. Sheoran, B. Giri, and W. S. Davidson. 2003. "Purification Strategies for Microbial Lipases." *Journal of Microbiological Methods* 52 (1):1–18.

Schrag, J. D., Y. Li, M. Cygler, D. Lang, T. Burgdorf, H.-J. Hecht, R. Schmid et al. 1997. "The Open Conformation of a *Pseudomonas* Lipase." *Structure* 5 (2):187–202.

Schuch, R., and K. D. Mukherjee. 1987. "Interesterification of Lipids Using an Immobilized Sn-1,3-Specific Triacylglycerol Lipase." *Journal of Agricultural and Food Chemistry* 35 (6):1005–1008.

Seitz, E. 1974. "Industrial Application of Microbial Lipases: A Review." *Journal of the American Oil Chemists' Society* 51 (2):12–16.

Sharma, R., Y. Chisti, and U. C. Banerjee. 2001. "Production, Purification, Characterization, and Applications of Lipases." *Biotechnology Advances* 19 (8):627–662.

Sharma, D., B. Sharma, and A. K. Shukla. 2011. "Biotechnological Approach of Microbial Lipase: A Review." *Biotechnology* 10:23–40.

Shimada, Y., C. Koga, A. Sugihara, T. Nagao, N. Takada, S. Tsunasawa, and Y. Tominaga. 1993. "Purification and Characterization of a Novel Solvent-Tolerant Lipase from *Fusarium Heterosporum*." *Journal of Fermentation and Bioengineering* 75 (5):349–352.

Sierra, G. 1957. "A Simple Method for the Detection of Lipolytic Activity of Micro-Organisms and Some Observations on the Influence of the Contact between Cells and Fatty Substrates." *Antonie van Leeuwenhoek* 23 (1):15–22.

Singh, A., and M. Mukhopadhyay. 2011. "Overview of Fungal Lipase: A Review." *Applied Biochemistry and Biotechnology* 166 (2):486–520.

Sztajer, H., and E. Zboińska. 1988. "Microbial Lipases in Biotechnology." *Acta Biotechnologica* 8 (2):169–175.

Taipa, M. A., M. R. Aires-Barros, and J. M. S. Cabral. 1992. "Purification of Lipases." *Journal of Biotechnology* 26 (2–3):111–142.

Tamio, M., M. Yuko, and M. Akira. 1995. "Purification and of *Penicillium Roqueforti* Iam 7268 Lipase." *Bioscience, Biotechnology, and Biochemistry* 59 (2):329–330.

Tsuzuki, W., and T. Suzuki. 1991. "Reactive Properties of the Organic Solvent-Soluble Lipase." *Biochimica et Biophysica Acta (BBA)—Lipids and Lipid Metabolism* 1083 (2):201–206.

Ulbert, O., K. Bélafi-Bakó, K. Tonova, and L. Gubicz. 2005. "Thermal Stability Enhancement of *Candida Rugosa* Lipase Using Ionic Liquids." *Biocatalysis and Biotransformation* 23 (3–4):177–183.

Uppenberg, J., S. Patkar, T. Bergfors, and T. A. Jones. 1994. "Crystallization and Preliminary X-Ray Studies of Lipase B from *Candida Antarctica*." *Journal of Molecular Biology* 235:790–792.

Van Camp, J., A. Huyghebaert, and P. Geoman. 1998. "Enzymatic Synthesis of Structured Modified Fats." In *Structural Modified Food Fats: Synthesis, Biochemistry, and Use*, edited by A. B. Christophe, 20–45. AOCS Press.

van Tilbeurgh, H., L. Sarda, R. Verger, and C. Cambillau. 1992. "Structure of the Pancreatic Lipase-Procolipase Complex." *Nature* 359 (6391):159–162.

Verger, R. 1997. "'Interfacial Activation' of Lipases: Facts and Artifacts." *Trends in Biotechnology* 15 (1):32–38.

Verger, R., and G. H. de Haas. 1976. "Interfacial Enzyme Kinetics of Lipolysis." *Annual Review of Biophysics and Bioengineering* 5:77–117.

Verger, R., M. C. E. Mieras, and G. H. de Haas. 1973. "Action of Phospholipase A at Interfaces." *Journal of Biological Chemistry* 248 (11):4023–4034.

Wang, Y., K. C. Srivastava, G. Shen, and H. Y. Wang. 1995. "Thermostable Alkaline Lipase from a Newly Isolated Thermophilic *Bacillus*, Strain A30-1 (ATCC 53841)." *Journal of Fermentation and Bioengineering* 79 (5):433–438.

Weete, J. D., L. Oi-Ming, and C. C. Akoh. 2008. "Microbial Lipases." In *Food Lipids: Chemistry, Nutrition, and Biotechnology*, 3rd ed., edited by Casmir C. Akoh and David B. Min. CRC Press/Taylor & Francis.

Westoby, M., J. Chrostowski, P. de Vilmorin, J. P. Smelko, and J. K. Romero. 2011. "Effects of Solution Environment on Mammalian Cell Fermentation Broth Properties: Enhanced Impurity Removal and Clarification Performance." *Biotechnology and Bioengineering* 108 (1):50–58.

Winkler, F. K., A. D'Arcy, and W. Hunziker. 1990. "Structure of Human Pancreatic Lipase." *Nature* 343 (6260):771–774.

Yahya, A. R. M., W. A. Anderson, and M. Moo-Young. 1998. "Ester Synthesis in Lipase-Catalyzed Reactions." *Enzyme and Microbial Technology* 23 (7–8):438–450.

Zhao, H., and W. Chen. 2008. "Chemical Biotechnology: Microbial Solutions to Global Change." *Current Opinion in Biotechnology* 19 (6):541–543.

3 Lipase Immobilization

The use of soluble and expensive enzymes on an industrial scale is still not feasible. From an economic point of view, it is crucial to recover, reuse, and retain the enzyme activity for continuous operations. The use of enzymes in solid form, via immobilization, may help in achieving this. However, when the enzyme is immobilized on a surface, the substrates and products diffusion to the enzyme active site may be restricted, which result in an additional add cost to the overall production process. The present chapter focuses on lipase immobilization and its advantages over soluble lipases and its effect on physiochemical characteristics, namely, their activity and stability. The different immobilization techniques are described, and examples of various immobilization methods are presented. In addition, different bioreactor configurations that use immobilized lipase are described.

3.1 LIPASE IMMOBILIZATION

Due to the globular protein nature of lipases, they are soluble in aqueous solutions. The uses of soluble lipases in industry are quite limited due to their natural instability under certain operating conditions, such as at high temperature and shear stress. In addition, the utilization of soluble lipases suffers from other processing difficulties, such as complexity in reusing the lipase and product contamination.

Various approaches that seek to overcome the problem of intimate contact between insoluble substrates and soluble lipase, which plays an important role in intervening lipase specificity, have been tested. These approaches include the generation of a stirred biphasic system, protein engineering, and lipase immobilization (Balcão et al., 1996; Klibanov et al., 1977). The use of organic solvent multiphase systems is useful when dissolving the reaction substrates, and hence enhancing reaction rates. This provides pseudohomogeneous solutions that are thermodynamically stable. Nevertheless, the practical use of lipase in pseudohomogeneous reaction solutions presents numerous technical and economic difficulties, such as product contamination from residual proteins, the use of lipase for a single reactor pass, and the requirements of a solvent recovery system. The most practical approach to solving these drawbacks is the use of lipase in an immobilized form.

The term *immobilized enzyme* was first coined in 1971 by Henniker (Hartmeier, 1988). By immobilization, the free movement of the lipase is restricted by localization in a defined region via a material, called a support, that is permeable to the substrates and product molecules. This has many advantages; the most important are stability enhancement, providing physical separation from the bulk reaction medium, and simplifying the recovery of the lipase (Cao, 2005; Iso et al., 2001). In addition, when the lipase is immobilized, it becomes independent within the reaction system, which extends its half-life (Balcão et al., 1996; Roig et al., 1987), facilitates reactor design, and controls the reaction (Hartmeier, 1985; Katchalski-Katzir, 1993). These features enable the cost-effective use of immobilized lipases for different applications (Kawakami et al., 2011; Rajendran et al., 2009; Tahoun, 1986; Wang et al., 2014).

Compared to soluble lipase, the activity of immobilized lipase may reduce due to the external or internal mass diffusion restrictions. Such effects lead to a reduction in system efficiency. The varying reduction levels depend on the properties and characteristics of the immobilized lipase. On the other hand, thermostability enhancement by immobilization allows the process to operate at higher temperatures, which can have a positive effect on the reaction rate and yield (Fjerbaek et al., 2009; Matsumoto and Ohashi, 2003), and, therefore, improve the overall process.

Although immobilization increases the manufacturing costs and underestimates the potential of immobilized lipase, this additional cost is lower than that associated with soluble lipase if cheaper supports are used. Typically, the high cost of a soluble lipase system arises from the cost of large numbers of enzyme (loading) and recovery unit requirements. In general, immobilized enzymes have to perform two essential functions: (1) noncatalytic functions designed to aid separation and (2) catalytic functions that convert substrates to product within a desired time and space (Cao, 2005).

3.2 SUPPORT MATERIALS

In immobilization, various carriers/supports have been used with enzymes. Porous glass and other derivatives (Cao et al., 1992; Marlot et al., 1985; Rucka and Turkiewicz 1990), activated carbon (Kandasamy et al., 2010), cellulose and other derivatives (Hwang et al., 2004), clay (de Oliveira et al., 2000; Lee and Akoh, 1998; Scherer et al., 2011), alumina (Bagi et al., 1997; Padmini et al., 1993), nylon (Pahujani et al., 2008), polyethylene and its derivatives (Watanabe et al., 1994), and polystyrene (Miletić et al., 2010) are the most common supports that have been used in lipase immobilization. The choice of a suitable support depends on its properties, such as its chemical composition, particle size, and surface area. These are essential for various potential applications (Villeneuve et al., 2000). Generally, porous supports are better than nonporous ones due to their larger surface area. However, they must have a structure that allows the lipase to bind and access the substrate with a minimum internal diffusion limitation. Further details on the effect of diffusion restrictions on reaction efficiency are discussed in Chapter 4.

3.3 METHODS OF IMMOBILIZATION

The immobilization of any enzyme can be carried out using different methods. These are classified into: (1) adsorption on hydrophobic and ionic exchange resins, (2) covalent attachment on highly activated supports, (3) entrapment, and (4) encapsulation in matrices, as shown in Figure 3.1. Combinations of two or more of these methods have also been developed (Katzbauer et al., 1995; Yadav and Jadhav, 2005; Zarcula et al., 2009). It has been demonstrated that such combinations can solve problems that cannot be solved using only one method.

The selection of the proper immobilization strategy is mainly dependent on the properties of the selected support and the lipase, where the interaction between these

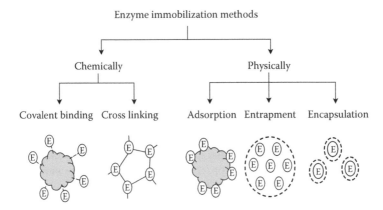

FIGURE 3.1 Enzyme immobilization methods.

two factors determines its mechanical, chemical, kinetic, and physical properties. Lipase tolerance to immobilization, microenvironment, functional groups on the protein surface, protein polarity, and substrate/product transport and other properties, such as the particle diameter, mechanical strength, and compression behavior, determines the reactor type (Hanefeld et al., 2009), have to be considered prior to commercialization.

3.3.1 ADSORPTION

Lipase immobilization via adsorption is the simplest and least expensive method, as it does not require any use of toxic chemicals. By adsorption, the ability to retain the specific activity and selectivity of the lipase, as compared to soluble lipase, remains unchanged even under harsher conditions (D'Souza, 1999). Therefore, it has been selected by most researchers. Among the immobilized lipases used in research and industry is *Candida antarctica* lipase B, which is immobilized by adsorption on a macroporous polyacrylate (Lewatit VP OC 1600) matrix. This immobilized form of lipase is known as Novozym®435, as it was developed by Novozymes.

The adsorption of the lipase onto an insoluble support is based on the attachment by noncovalent linkages, such as van der Waals, hydrogen bonds, or ionic interactions without any pre-activation step. The interaction mainly depends on the existing surface properties of the support and on the type of lipase amino acids exposed on the surface. Hydrophobic supports are most commonly used for lipase immobilization by adsorption (Brady et al., 1988; de Oliveira et al., 2000). Palomo et al. (2002a) tested several hydrophobic support materials, including octylagarose and octadecyl-Sepabeads, for lipase immobilization and found that the activity and enantioselectivity of such derivatives were much higher than that of free lipase. These hydrophobic supports also have a higher activity than hydrophilic supports. This has been proven by G. J. Chen et al. (2012) who compared the efficiency of lipase immobilized on three membranes with different hydrophobic/hydrophilic properties. The lipase immobilized on a hydrophobic membrane exhibited a more than 11-fold increase in activity compared to that immobilized on a hydrophilic membrane.

Generally, adsorption can be performed by dissolving the lipase in a buffer solution and mixing it with the support material, which is an adsorbent under properly controlled conditions. This is followed by washing with a buffer solution after sufficient incubation and the removal of the supernatant solution (Yong et al., 2010). The process initially diffuses the lipase from the bulk solution to the support surface and then binds the adsorbed lipase to the adsorption sites. The rate of the later step is usually much higher than the former, therefore the overall process is diffusion controlled. The adsorption process depends on many factors, such as pH, protein properties, and the nature of the support surface. A little shift in pH may affect the interaction and lead to the release of the lipase from the support. Due to the weak interaction and the marginal effect on lipase conformation, regeneration of the inactive enzyme by adding a fresh enzyme is possible (Palomo et al., 2002a; Tsai and Shaw, 1998). Although this technique is cheap, leaching remains one of its most significant shortcomings.

The Langmuir isotherm, shown in Equation 3.1, is characterized by a sharp and steep slope at a lower concentration range, and a plateau at a higher concentration. It is a classic model of adsorption. In this model, the adsorbed lipase is almost equivalent to a complete monolayer.

$$C_{lip.ads.} = \frac{C_{lip.ads.max} K_{lip.} C_{lip.free}}{1 + K_{lip.} C_{lip.free}} \qquad (3.1)$$

where $C_{lip.ads.}$ and $C_{lip.free}$ represent the concentration of lipase adsorbed in the bulk of the supernatant solution, respectively. $K_{lip.}$ represents the adsorption equilibrium constant and $C_{lip.ads.max}$ represents the maximum concentration of lipase that can be adsorbed per unit area. This is also temperature dependent, where $K_{lip.}$ can be expressed by the van't Hoff's relationship in term of the equilibrium constant, as shown in Equation 3.2. The maximum concentration of the temperature dependence is usually represented as in Equation 3.3.

$$K_{lip.} = \beta \exp\left[\frac{-\Delta h_{lip.ads.}}{RT}\right] \qquad (3.2)$$

$$C_{lip.ads.max} = \alpha(1 + \varepsilon T) \qquad (3.3)$$

where T is the absolute temperature; $\Delta h_{lip.ads.}$ is the standard enthalpy change associated with the adsorption of lipase onto the membrane surface; α, ε, and β are constants; and R is the ideal gas constant.

Although, this model describes enzyme adsorption, other studies have shown that the lipase distribution on the support surface can be presented by a Freundlich model (Knezevic et al., 1998; Mojovic et al., 1998) that describes nonidentical adsorption on the support surface, or by a Redlich-Peterson model (Al-Duri and Yong, 2000) that combines the Langmuir and Freundlich isotherms. The Langmuir model reflects irreversible adsorption and is based on the assumption that every adsorption site is identical and energetically equivalent. However, this is a condition that is rarely

TABLE 3.1

Parameters of the Langmuir Adsorption Isotherm Model Coupled with the Van't Hoff's Relationship Model

Parameter	Unit	Estimated Value
A	$LU\,g^{-1}$	3264.7
ε	K^{-1}	3.4×10^{-3}
$-\Delta h_{lip.ads.}/R$	K	6668.9
	$ml\,LU^{-1}$	4.2×10^{6}

fulfilled (Andrade and Hlady, 1986). Nevertheless, many authors have employed the Langmuir model to correlate experimental data in order to compare different lipase/adsorbent systems (Akova and Ustun, 2000; Mendieta-Taboada et al., 2001; Yong et al., 2010). Shamel et al. (2005) investigated the temperature effect on the adsorption of *Mucor miehei* lipase. Their model constants are shown in Table 3.1.

It has been observed that not all porous supports are suitable for immobilization, owing to the limitations of pore size, which should be at least the same as that of the lipase. By studying the adsorption of *Rhizomucor miehei* lipase on a porous inorganic support, it was found that the activity depended on a pore size in the range of 100 nm. Beyond this value, the activity is independent suggesting that the internal diffusion restriction is insignificant (Bosley and Clayton, 1994). Similar observations have also been noted with the immobilization of *Candida rugosa* lipase on different hydrophobic carriers of various pore sizes (Al-Duri et al., 1995).

3.3.2 COVALENT BINDING

The covalent attachment of enzymes on supports is another popular immobilization technique where enzymes are retained on the support surface by forming a strong covalent bonds. It is a chemical modification of the enzyme's amino acids by binding the proteins, known as cross-linking, or between the proteins' amino acid residues and the support. In cross-linking, the enzyme serves as the carrier material. In this method, functional groups in the support usually have to be activated by activators such as cyanogen bromide (Axen et al., 1967), cyanuric chloride (Miletić et al., 2009), and sodium periodate (Girelli et al., 2012) in the case of simple covalent binding; or by bi-/multifunctional reagents, such as glutaraldehyde (Chen et al., 2008), hexamethylene diisocyanate (Erdemir and Yilmaz, 2009; Ozmen et al., 2009), and toluene diisocyanate (Kaar, 2011) to make the functional groups electrophilic. The latter reacts with strong nucleophiles to form a covalent bond (Walker and Rapley, 2009).

The main advantage of this method is high operational stability as a result of the interaction with a low leakage from the lipase and the support even if the process temperature, pH, or solvent is changed. This is a crucial feature in the feasibility of any industrial process. Hence, it is considered better than adsorption, as diffusion restrictions are fewer and the bond strength prevents the enzyme from leaking into the solution. Rodrigues et al. (2002) studied the covalent immobilization of *Chromobacterium*

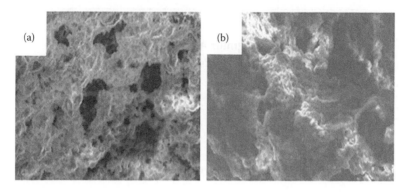

FIGURE 3.2 Electron microscope images of (a) surface-modified mesoporous activated carbon and (b) lipase immobilized on surface-modified mesoporous activated carbon. (From Ramani, K., R. Boopathy, A. B. Mandal, and G. Sekaran, 2011, *Catalysis Communications* 14 (1):82–88. With permission.)

viscosum lipase on Eudragit S-100 by activating the polymer with carbodiimide and reported that this derivative can be effectively used for the esterification of different alcohols and fatty acids in both organic and two-phase systems. Figure 3.2 shows an electron microscope photograph of lipase produced from *Pseudomonas gessardii*, immobilized covalently on a surface of modified mesoporous activated carbon using ethylenediamine and glutaraldehyde (Ramani et al., 2011). Although this method is advantageous, its shortcoming is a relatively low initial activity due to the potential blockage of some active sites thanks to immobilization.

3.3.3 ENTRAPMENT

The entrapment process involves physically entrapping the enzyme with insoluble polymers of synthetic or natural origin. It is a type of gel entrapment in which the lipase is brought into a monomeric solution, which upon polymerization leads to its entrapment. By using this method, the lipase is kept free in solution; however, it is also restricted by the polymer (Murty et al., 2002). The advantage of such a technique is that it is simple, the enzyme does not interact with the polymer, denaturation is avoided, and it has a high initial yield. However, the mass transfer limitation, in the form of internal diffusion resistance, is a problem, as the substrate/product diffusion rate across the membrane is a limiting factor (Villeneuve et al., 2000).

Among the common matrices used for entrapment is a polyacrylamide gel that has been used with the extracellular lipase from *Penicillium chrysogenum* (Shafei and Allam, 2010). This gel is not suitable for use in food applications due to its toxicity (D'Souza, 1999).

3.3.4 ENCAPSULATION

Microencapsulation is a similar technique to entrapment but less common. In this method, a membrane-like support is formed around the enzyme. Therefore, both the lipase and its environment are immobilized. Encapsulation can be carried out using polymers, such as alginate, acrylic polymers, hydrogels, and microemulsion based gels (Yadav and Jadhav, 2005). *C. antarctica* lipase has been encapsulated in hydrophobic sol-gel materials, and nanoparticles of iron oxide have been used in the encapsulation to facilitate separation (Reetz et al., 1996a). Yang et al. (2009) immobilized *Arthrobacter* sp. lipase via encapsulation in hydrophobic sol-gel materials and in their study the overall activity was approximately 14-fold that of the soluble lipase. In addition, the immobilized lipase showed higher thermal and operational stabilities. Moreover, the activity of the encapsulated *C. rugosa* lipase was found to be higher than with a covalently immobilized lipase (Yilmaz and Sezgin, 2012).

3.3.5 COMBINATION OF METHODS

The drawbacks of previous methods can be overcome by a combining two of any immobilization methods (Kanwar et al., 2004; Yadav and Jadhav, 2005; Zarcula et al., 2009). Ursoiu et al. (2011) deposited immobilized *C. antarctica* lipase B via sol-gel entrapment on support material. Yadav and Jadhav (2005) tested the preimmobilization of *C. antarctica* lipase B on hexagonal mesoporous silica using physical adsorption, followed by encapsulation in calcium alginate beads, which resulted in a reusable lipase with no leaching even after the fourth reuse.

3.4 LIPASE PHYSIOCHEMICAL PROPERTIES IMPROVEMENT

As noted earlier, immobilization is not only intended to retain lipase activity for reuse, but has also been proposed to improve the poor percentage of native lipases. Thus, when a certain immobilization technique is adopted, improvements in the performance of the lipase should be a factor. The enhancement should be in the main catalytic characteristics.

3.4.1 STABILITY

The stability of lipase is one of its prominent properties that needs to be considered for improvement via immobilization. The long-term storage and thermal stability of any immobilized enzyme is dependent upon the (1) lipase interaction with the support, (2) binding position, (3) microenvironment, (4) chemical and physical structure of the support, (5) the properties of the spacer that links the lipase to the support, and (6) immobilization conditions. Hence, random immobilizations of the lipase usually do not improve the stability (Chibata et al., 1986; Hartmeier, 1985; Martinek et al., 1977).

With lipase in immobilized form, it will not be in direct contact with any external hydrophobic interface that may inactivate soluble proteins. For example, in the presence of an organic solvent the immobilized lipase may be in contact but will not be soluble, thus preventing inactivation from occurring. This was investigated by Nawani et al. (2006), who compared the catalytic properties of *Bacillus* sp. lipases immobilized by different techniques. In their study, stability was improved with an optimum temperature of 5°C higher than that of the soluble lipase. This was reported after immobilization by adsorption on silica and HP-20 beads followed by crosslinking with gluteraldehyde on HP-20. Montero et al. (1993) compared the stability of *C. rugosa* lipase immobilized on microporous polypropylene with native lipase. The free lipase showed maximal activity at 37°C, whereas for immobilized lipase it was around 45°C. Similar results were observed for *C. antarctica* lipase B when immobilized covalently on sepharose, alumina, and silica (Arroyo et al., 1999). It was also reported that lipases immobilized covalently are more stable than those immobilized physically. The covalent immobilization of *C. rugosa* on hydrophilic magnetic microspheres was also investigated by Yong et al. (2008), where better resistance to pH changes and temperature was reported in comparison to free lipase. The improved thermal stability could be due to either the introduction of intramolecular bonds between the lipase and the support during immobilization, or the prevention of conformational inactivation of the lipase by immobilization. The wider pH range depends on the ionic change environment around the active site of the lipase during immobilization (Krajewska et al., 1990). Polyanionic supports shift the optimum pH toward an alkaline direction (high pH), whereas the polycationic shifts toward an acid direction (low pH).

3.4.2   ACTIVITY

As mentioned in Chapter 2, lipases show two different conformations: closed form, where the active sites are inaccessible to the reaction medium; and open form, where the lid is displaced and the active sites are exposed to the reaction medium. Both forms are in equilibrium. When in the presence of substrate droplets, lipase becomes adsorbed on the droplet interface and the conformation is shifted toward the open form. Generally, lipase activity enhancement can be achieved by (1) microenvironment effect; (2) conformational change; (3) reducing the diffusion effect; (4) reducing water partition effects, especially in organic solvent; or (5) the binding mode.

Activity enhancement in organic solvents has been demonstrated after immobilization with sol-gel and a significant increase in activity was reported (Reetz et al., 2000). Reetz et al. (1996b) entrapped lipases in a nanoporous hydrophobic sol-gel in the presence of porous glass beads and reported that the activity of the lipases was enhanced by 88-fold with higher activity recovery. Goto et al. (2005) found that lipase entrapped in an n-vinyl-2-pyrrolidone gel matrix was 51-fold more active than the native enzyme. However, if the substrate is large or hydrophilic, the presence of a hydrophobic support may generate some stearic hindrances that reduce lipase activity (Villeneuve et al., 2000) and vice versa (Yahya et al., 1998). This in mainly because of the blockage of the active site to which large substrates have no access.

3.4.3   SELECTIVITY

Selectivity improvement is a critical requirement for industrial applications of lipase. This includes substrate selectivity, stereoselectivity, regioselectivity, and enantioselectivity. Adsorption of *Candida rugosa* on celite was reported to enhance the stability of lipase and improve its enantioselectivity up to 3-fold (Ogino, 1970). Entrapment in cellulose acetate–TiO_2 gel fiber improved the selectivity of *Rhizomucor miehei* lipase in the hydrolysis of 1,2-diacetoxypropane (Ikeda and Kurokawa, 2001). Also the enantioselectivity of pegylated *P. cepacea lipase* was increased 3-fold by entrapment in Ca-alginate gel beads (Palomo et al., 2003). Physical immobilization of *C. antarctica* lipase B by adsorption onto octadecyl-Sepabeads (hydrophobic support) did not show any appreciable enantioselectivity. However, the covalent immobilization onto the glutaraldehyde (hydrophilic support) derivative showed high enantioselectivity (Palomo et al., 2002b). This is mainly because glutaraldehyde is soluble in an aqueous media and can form inter- and intrabonds. If immobilized, it alters the rigidity of the lipase resulting in a conformational change from closed form to open form.

3.5 IMMOBILIZED LIPASE REACTORS

Enzymatic systems have been suggested to reduce the obstacles associated with the use of chemical catalysts. It is essential for such systems to be cost-effective with high product yields. Several bioreactor configurations, such as batch stirred tank reactors (BSTRs), packed bed reactors (PBRs), continuous stirred tank reactors (CSTRs), fluidized bed reactors (FBRs), and membrane reactors (MRs), have been proposed in using immobilized lipase. The continuous operation of these reactors with good process control and good contact between reaction components and the immobilized lipase are the main factors that determine process and system success. The contact between the reaction mixture and immobilized lipase can be improved by enhancing the convection. However, the convection can be associated with high shear rates or high surface tension, which can lead to faster lipase deactivation.

3.5.1 BATCH STIRRED TANK REACTORS

Bioreactors that operate in batchwise are the most commonly used. They consist of a vessel in which the reaction mixture is stirred by mechanical means, such as impellers or magnetic bars. Initially, the bioreactor is filled with a reaction medium containing the substrate and the operating conditions are adjusted. Then the enzyme is added and the reaction is left to proceed until the desired conversion takes place. This step is followed by immobilized lipase separation from the reaction medium via filtration or centrifugation and washing in a solvent before reuse. Due to economic concerns associated with emptying, cleaning, and refilling the reactor, large-scale use of batch reactors is limited. These types of bioreactors are used for kinetic reaction studies.

3.5.2 PACKED (FIXED) BED REACTORS

A packed bed reactor is a continuous system with a constant flow rate of reaction medium fed to the bioreactor. Due to their high efficiency, low cost, and easy operation and maintenance, packed bed reactors have been used on a large scale. Figure 3.3 shows a schematic diagram of a packed bed reactor used for the production of structured triacylglycerols in the presence of immobilized lipase. Immobilized lipases are usually packed forming a bed with a large surface area per unit volume. The reaction mixture is kept in a reservoir under controlled conditions and pumped through the bed. As they pass through the bed, the substrates are converted into the desired products, and the degree of conversion is controlled by the residence time

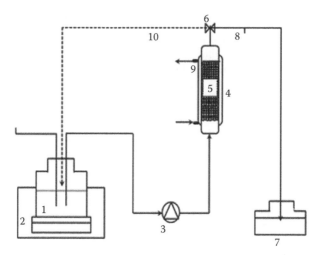

FIGURE 3.3 Schematic diagram of immobilzed lipase PBR: 1, substrate reservoir; 2, temperature control; 3, peristaltic pump; 4, water jacket; 5, bed of immobilized enzyme; 6, three-way valve; 7, product reservoir; 8, sampling point; 9, cooling/heating water; 10, recirculation. (From Hita, E., A. Robles, B. Camacho, P. A. González, L. Esteban, M. J. Jiménez, M. M. Muñío, and E. Molina, 2009, *Biochemical Engineering Journal* 46 (3):257–264. With permission.)

determined by the bed height and the flow rate of the reaction medium. The reactor can be fed from the bottom or the top. On a small scale, it is more common to feed the substrates from the bottom, as it is easier to maintain the liquid level above the lipase bed. However, on a large scale, top feeding is frequently used to take advantage of gravitational forces that can reduce the energy required for pumping the liquid. Such a configuration is applicable within biphasic systems. However, countercurrent flows can also be used, where the two phases may be pumped in opposite directions with the densest flowing downward. There are two limitations that must be considered when using a packed bed reactor. These are the intraparticle diffusion restrictions on reaction rates and the high pressure drop across reactor packing. External limitations are generally overcome by increasing substrate flow rates through the bioreactor. The use of small particle sizes reduces the effect of internal diffusion but may increase the drop in pressure and cause channeling.

3.5.3 Continuous Stirred Tank Reactors

In a continuous stirred tank reactor, a constant flow of reaction substrates is fed to the reactor, where the immobilized lipase is suspended in an agitated vessel. The main character of this type of reaction is that there is no temperature or concentration gradients due to efficient mixing that promotes intimate contact of the reaction mixture with the immobilized lipase. Like batchwise reactors, immobilized lipase can be retained within the bioreactor by filtration. This is known to have lower construction cost. However, it requires larger volumes than a PFR to achieve the same reaction. Commonly, a microfilter is provided at the bioreactor outlet to prevent immobilized lipase from leaving the reactor.

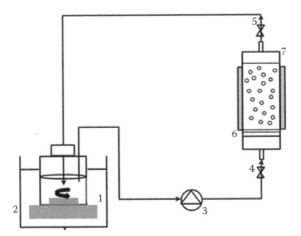

FIGURE 3.4 Experimental setup for lipase-catalyzed esterification in an FBR system: 1, reaction mixture reservoir; 2, temperature control; 3, peristaltic pump; 4, sampling valve (inlet side); 5, sampling valve (outlet side); 6, FBR with the immobilized lipase; 7, water jacket with cooling/heating water. (Redrawn from Hajar, M., and F. Vahabzadeh, 2014, *Industrial Crops and Products* 59:252–259. With permission.)

3.5.4 FLUIDIZED BED REACTORS

The fluidized bed reactor is a hybrid of a CSTR and a PBR, where the reaction fluid velocity is higher than the fluidization velocity of the bed. In a fluidized bed reactor, substrates are passed upward through the immobilized lipase at a velocity that lifts the solid particles, yet at the same time it does not sweep them away. Such reactors have the advantage of eliminating the channeling problem with a smaller drop in pressure compared to PBRs. However, fluidized bed reactors have only been used with low lipase concentrations, as a large void volume is require to keep the enzyme suspended. Figure 3.4 shows an example of an experimental setup for Novozym® 435 lipase for biolubricant production from castor oil in an FBR system (Hajar and Vahabzadeh, 2014).

3.5.5 MEMBRANE REACTORS

The use of membrane reactors offers several advantages, including (1) a high specific surface area, (2) instantaneous reaction and separation of substrate and product, (3) reusing the lipase, and (4) continuous operation. Immobilization of lipase onto a membrane offers many advantages, such as a low drop in pressure when continuously operated, and high operational stability with low external and internal diffusional resistance (Giorno and Drioli, 2000). In these bioreactors, the lipase is immobilized on a membrane, which may be either a flat sheet (FSMR) or hollow fiber (HFMR).

A variety of membrane materials (hydrophobic and hydrophilic) and reactor configurations have been studied. Hydrophobic membrane reactors are made of polypropylene (Atia et al., 2003; Malcata, 1992a; Malcata et al., 1991), polytetrafluoroethylene

(Atia et al., 2003; Goto et al., 1992), polyvinyl chloride (Rucka and Turkiewicz, 1990; Shaw et al., 1990), poly(vinylidene difluoride) (Tsai and Shaw, 1998), or polyetherimide (Merçon et al., 1997), where the lipase can be immobilized on the aqueous side of the membrane. Conversely, in hydrophilic membrane reactors, lipase is immobilized on the organic side of the membrane, which is made of cellulose (Guit et al., 1991;

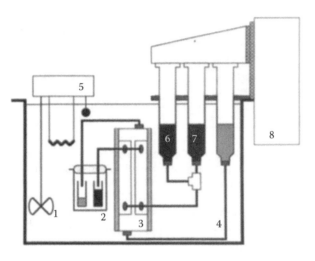

FIGURE 3.5 Schematic diagram of flat sheet bioreactor apparatus used for butteroil hydrolysis: 1, stirrer; 2, products; 3, membrane reactor; 4, water bath; 5, temperature controller; 6, oil; 7, buffer; 8, pump. (From Malcata, F. X., C. G. Hill, and C. H. Amundson, 1991, *Biotechnology and Bioengineering* 38 (8):853–868. With permission.)

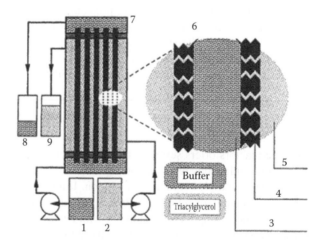

FIGURE 3.6 Schematic diagram of a membrane reactor with lipase-coated hollow fiber: 1, buffer solution; 2, triglycerides; 3, lumen; 4, hollow fiber wall; 5, shell side; 6, lipase coated hollow fiber; 7, hollow fiber reactor; 8, product 1 collection; 9, product 2 collection. (From Malcata, F. X., C. G. Hill, and C. H. Amundson, 1992, *Biotechnology and Bioengineering* 39 (10):1002–1012. With permission.)

Pronk et al., 1988), acryl (Taylor et al., 1986), or polyamide (Giorno et al., 1997). *Aspergillus niger* lipase immobilized by adsorption on a flat sheet of microporous polypropylene membrane for continuous hydrolysis of butter fat glycerides (Malcata et al., 1991) has been used. Figure 3.5 shows a schematic diagram of an example of FSMR in use.

Hollow fiber membrane reactors are well known to provide the highest surface-to-volume ratio (Heath and Belfort, 1990). A typical system, as shown in Figure 3.6, consists of a bundle of hollow fibers that are made of hydrophobic microporous poly-propylene. A tangential flow membrane has been reported by many investigators and was found to yield good results for lipase catalyzed hydrolysis and ester synthesis (Hoq et al., 1985).

REFERENCES

Akova, A., and G. Ustun. 2000. "Activity and Adsorption of Lipase from Nigella Sativa Seeds on Celite at Different pH Values." *Biotechnology Letters* 22 (5):355–359.

Al-Duri, B., and Y. P. Yong. 2000. "Lipase Immobilisation: An Equilibrium Study of Lipases Immobilised on Hydrophobic and Hydrophilic/Hydrophobic Supports." *Biochemical Engineering Journal* 4 (3):207–215.

Al-Duri, B., E. Robinson, S. McNerlan, and P. Bailie. 1995. "Hydrolysis of Edible Oils by Lipases Immobilized on Hydrophobic Supports: Effects of Internal Support Structure." *Journal of the American Oil Chemists' Society* 72 (11):1351–1359.

Andrade, J., and V. Hlady. 1986. "Protein Adsorption and Materials Biocompatibility: A Tutorial Review and Suggested Hypotheses." In *Biopolymers/Non-Exclusion HPLC*, Advances in Polymer Science, vol. 79, 1–63. Springer-Verlag.

Arroyo, M., J. M. Sánchez-Montero, and J. V. Sinisterra. 1999. "Thermal Stabilization of Immobilized Lipase B from Candida Antarctica on Different Supports: Effect of Water Activity on Enzymatic Activity in Organic Media." *Enzyme and Microbial Technology* 24 (1–2):3–12.

Atia, K. S., M. B. El-Arnaouty, S. A. Ismail, and A. M. Dessouki. 2003. "Characterization and Application of Immobilized Lipase Enzyme on Different Radiation Grafted Polymeric Films: Assessment of the Immobilization Process Using Spectroscopic Analysis." *Journal of Applied Polymer Science* 90 (1):155–167.

Axen, R., J. Porath, and S. Ernback. 1967. "Chemical Coupling of Peptides and Proteins to Polysaccharides by Means of Cyanogen Halides." *Nature* 214 (5095):1302–1304.

Bagi, K., L. M. Simon, and B. Szajáni. 1997. "Immobilization and Characterization of Porcine Pancreas Lipase." *Enzyme and Microbial Technology* 20 (7):531–535.

Balcão, V. M., A. L. Paiva, and F. Xavier Malcata. 1996. "Bioreactors with Immobilized Lipases: State of the Art." *Enzyme and Microbial Technology* 18 (6):392–416.

Bosley, J. A., and J. C. Clayton. 1994. "Blueprint for a Lipase Support: Use of Hydrophobic Controlled-Pore Glasses as Model Systems." *Biotechnology and Bioengineering* 43 (10):934–938.

Brady, C., L. Metcalfe, D. Slaboszewski, and D. Frank. 1988. "Lipase Immobilized on a Hydrophobic, Microporous Support for the Hydrolysis of Fats." *Journal of the American Oil Chemists' Society* 65 (6):917–921.

Cao, L. 2005. "Immobilised Enzymes: Science or Art?" *Current Opinion in Chemical Biology* 9 (2):217–226.

Cao, S., Z. Liu, Y. Feng, L. Ma, T. Z. Ding, and Y. H. Cheng. 1992. "Esterification and Transesterification with Immobilized Lipase in Organic Solvent." *Applied Biochemistry and Biotechnology* 32 (1):1–6.

Chen, B., J. Hu, E. M. Miller, W. Xie, M. Cai, and R. A. Gross. 2008. "Candida Antarctica Lipase B Chemically Immobilized on Epoxy-Activated Micro- and Nanobeads: Catalysts for Polyester Synthesis." *Biomacromolecules* 9 (2):463–471.

Chen, G.-J., C.-H. Kuo, C.-I. Chen, C.-C. Yu, C.-J. Shieh, and Y.-C. Liu. 2012. "Effect of Membranes with Various Hydrophobic/Hydrophilic Properties on Lipase Immobilized Activity and Stability." *Journal of Bioscience and Bioengineering* 113 (2):166–172.

Chibata, I., T. Tosa, and T. Sato. 1986. "Biocatalysis: Immobilized Cells and Enzymes." *Journal of Molecular Catalysis* 37 (1):1–24.

D'Souza, S. F. 1999. "Immobilized Enzymes in Bioprocess." *Current Science* 77 (1):69–80.

de Oliveira, P. C., G. M. Alves, and H. F. de Castro. 2000. "Immobilisation Studies and Catalytic Properties of Microbial Lipase onto Styrene-Divinylbenzene Copolymer." *Biochemical Engineering Journal* 5 (1):63–71.

Erdemir, S., and M. Yilmaz. 2009. "Synthesis of Calix[N]Arene-Based Silica Polymers for Lipase Immobilization." *Journal of Molecular Catalysis B: Enzymatic* 58 (1–4): 29–35.

Fjerbaek, L., K. V. Christensen, and B. Norddahl. 2009. "A Review of the Current State of Biodiesel Production Using Enzymatic Transesterification." *Biotechnology and Bioengineering* 102 (5):1298–1315.

Giorno, L., and E. Drioli. 2000. "Biocatalytic Membrane Reactors: Applications and Perspectives." *Trends in Biotechnology* 18 (8):339–349.

Giorno, L., R. Molinari, M. Natoli, and E. Drioli. 1997. "Hydrolysis and Regioselective Transesterification Catalyzed by Immobilized Lipases in Membrane Bioreactors." *Journal of Membrane Science* 125 (1):177–187.

Girelli, A. M., L. Salvagni, and A. M. Tarola. 2012. "Use of Lipase Immobilized on Celluse Support for Cleaning Aged Oil Layers." *Journal of the Brazilian Chemical Society* 23:585–592.

Goto, M., M. Goto, F. Nakashio, K. Yoshizuka, and K. Inoue. 1992. "Hydrolysis of Triolein by Lipase in a Hollow Fiber Reactor." *Journal of Membrane Science* 74 (3):207–214.

Goto, M., C. Hatanaka, and M. Goto. 2005. "Immobilization of Surfactant–Lipase Complexes and Their High Heat Resistance in Organic Media." *Biochemical Engineering Journal* 24 (1):91–94.

Guit, R. P. M., M. Kloosterman, G. W. Meindersma, M. Mayer, and E. M Meijer. 1991. "Lipase Kinetics: Hydrolysis of Triacetin by Lipase from Candida Cylindracea in a Hollow-Fiber Membrane Reactor." *Biotechnology and Bioengineering* 38 (7):727–732.

Hajar, M., and F. Vahabzadeh. 2014. "Modeling the Kinetics of Biolubricant Production from Castor Oil Using Novozym 435 in a Fluidized-Bed Reactor." *Industrial Crops and Products* 59:252–259.

Hanefeld, U., L. Gardossi, and E. Magner. 2009. "Understanding Enzyme Immobilisation." *Chemical Society Reviews* 38 (2):453–468.

Hartmeier, W. 1985. "Immobilized Biocatalysts—From Simple to Complex Systems." *Trends in Biotechnology* 3 (6):149–153.

Hartmeier, W. 1988. *Immobilized Biocatalysts: An Introduction.* Springer-Verlag.

Heath, C., and G. Belfort. 1990. "Membranes and Bioreactors." *International Journal of Biochemistry* 22 (8):823–835.

Hita, E., A. Robles, B. Camacho, P. A. González, L. Esteban, M. J. Jiménez, M. M. Muñío, and E. Molina. 2009. "Production of Structured Triacylglycerols by Acidolysis Catalyzed by Lipases Immobilized in a Packed Bed Reactor." *Biochemical Engineering Journal* 46 (3):257–264.

Hoq, M. M., H. Tagami, T. Yamane, and S. Shimizu. 1985. "Some Characteristics of Continuous Glycerides Synthesis by Lipase in Microporous Hydrophobic Membrane Bioreactor." *Agricultural and Biological Chemistry* 49 (2):335–342.

Hwang, S., J. Ahn, S. Lee, T. Gyu Lee, S. Haam, K. Lee, I.-S. Ahn, and J.-K. Jung. 2004. "Evaluation of Cellulose-Binding Domain Fused to a Lipase for the Lipase Immobilization." *Biotechnology Letters* 26 (7):603–605.

Ikeda, Y., and Y. Kurokawa. 2001. "Hydrolysis of 1,2-Diacetoxypropane by Immobilized Lipase on Cellulose Acetate-Tio2 Gel Fiber Derived from the Sol-Gel Method." *Journal of Sol-Gel Science and Technology* 21 (3):221–226.

Iso, M., B. Chen, M. Eguchi, T. Kudo, and S. Shrestha. 2001. "Production of Biodiesel Fuel from Triglycerides and Alcohol Using Immobilized Lipase." *Journal of Molecular Catalysis B: Enzymatic* 16 (1):53–58.

Kaar, J. L. 2011. "Lipase Activation and Stabilization in Room-Temperature Ionic Liquids." In *Enzyme Stabilization and Immobilization*, edited by S. D. Minteer, 25–35. Human Press.

Kandasamy, R., L. J. Kennedy, C. Vidya, R. Boopathy, and G. Sekaran. 2010. "Immobilization of Acidic Lipase Derived from Pseudomonas Gessardii onto Mesoporous Activated Carbon for the Hydrolysis of Olive Oil." *Journal of Molecular Catalysis B: Enzymatic* 62 (1):58–65.

Kanwar, S. S., M. Srivastava, S. S. Chimni, I. A. Ghazi, R. K. Kaushal, and G. K. Joshi. 2004. "Properties of an Immobilized Lipase of *Bacillus Coagulans* BTS-1." *Acta Microbiologica et Immunologica Hungarica* 51 (1):57–73.

Katchalski-Katzir, E. 1993. "Immobilized Enzymes: Learning from Past Successes and Failures." *Trends in Biotechnology* 11 (11):471-478.

Katzbauer, B., M. Narodoslawsky, and A. Moser. 1995. "Classification System for Immobilization Techniques." *Bioprocess and Biosystems Engineering* 12 (4):173–179.

Kawakami, K., Y. Oda, and R. Takahashi. 2011. "Application of a Burkholderia Cepacia Lipase-Immobilized Silica Monolith to Batch and Continuous Biodiesel Production with a Stoichiometric Mixture of Methanol and Crude Jatropha Oil." *Biotechnology for Biofuels* 4 (1):42.

Klibanov, A. M., G. P. Samokhin, K. Martinek, and I. V. Berezin. 1977. "A New Approach to Preparative Enzymatic Synthesis." *Biotechnology and Bioengineering* 19 (9):1351–1361.

Knezevic, Z., L. Mojovic, and B. Adnadjevic. 1998. "Palm Oil Hydrolysis by Lipase from *Candida Cylindracea* Immobilized on Zeolite Type Y." *Enzyme and Microbial Technology* 22 (4):275–280.

Krajewska, B., M. Leszko, and W. Zaborska. 1990. "Urease Immobilized on Chitosan Membrane: Preparation and Properties." *Journal of Chemical Technology and Biotechnology* 48:337–350.

Lee, K.-T., and C. C. Akoh. 1998. "Immobilization of Lipases on Clay, Celite 545, Diethy laminoethyl-, and Carboxymethyl-Sephadex and Their Interesterification Activity." *Biotechnology Techniques* 12 (5):381–384.

Malcata, F. X., C. G. Hill, and C. H. Amundson. 1991. "Use of a Lipase Immobilized in a Membrane Reactor to Hydrolyze the Glycerides of Butteroil." *Biotechnology and Bioengineering* 38 (8):853–868.

Malcata, F. X., H. S. Garcia, C. G. Hill, and C. H. Amundson. 1992a. "Hydrolysis of Butteroil by Immobilized Lipase Using a Hollow-Fiber Reactor: Part I. Lipase Adsorption Studies." *Biotechnology and Bioengineering* 39 (6):647–657.

Malcata, F. X., C. G. Hill, and C. H. Amundson. 1992b. "Hydrolysis of Butteroil by Immobilized Lipase Using a Hollow-Fiber Reactor: Part III. Multiresponse Kinetic Studies." *Biotechnology and Bioengineering* 39 (10):1002–1012.

Marlot, C., G. Langrand, C. Triantaphylides, and J. Baratti. 1985. "Ester Synthesis in Organic Solvent Catalyzed by Lipases Immobilized on Hydrophilic Supports." *Biotechnology Letters* 7 (9):647–650.

Martinek, K., A. M. Klibanov, V. S. Goldmacher, and I. V. Berezin. 1977. "The Principles of Enzyme Stabilisation. I. Increase in Thermostability of Enzymes Covalently Bound to a Complementary Surface of a Polymer Support in a Multi-Point Fashion." *Biochim Biophys Acta* 485:1–12.

Matsumoto, M., and K. Ohashi. 2003. "Effect of Immobilization on Thermostability of Lipase from *Candida Rugosa*." *Biochemical Engineering Journal* 14 (1):75–77.

Mendieta-Taboada, O., E. S. Kamimura, and F. Maugeri. 2001. "Modelling and Simulation of the Adsorption of the Lipase from *Geotrichum* Sp. on Hydrophobic Interaction Columns." *Biotechnology Letters* 23 (10):781–786.

Merçon, F., V. L. Erbes, G. L. Sant'Anna Jr., and R. Nobrega. 1997. "Lipase Immobilized Membrane Reactor Applied to Babassu Oil Hydrolysis." *Brazilian Journal of Chemical Engineering* 14 (1).

Miletić, N., R. Rohandi, Z. Vuković, A. Nastasović, and K. Loos. 2009. "Surface Modification of Macroporous Poly(Glycidyl Methacrylate-Co-Ethylene Glycol Dimethacrylate) Resins for Improved *Candida Antarctica* Lipase B Immobilization." *Reactive and Functional Polymers* 69 (1):68–75.

Miletić, N., V. Abetz, K. Ebert, and K. Loos. 2010. "Immobilization of *Candida Antarctica* Lipase B on Polystyrene Nanoparticles." *Macromolecular Rapid Communications* 31 (1):71–74.

Mojovic, L., Z. Knezevic, R. Popadic, and S. Jovanovic. 1998. "Immobilization of Lipase from *Candida Rugosa* on a Polymer Support." *Applied Microbiology and Biotechnology* 50 (6):676–681.

Montero, S., A. Blanco, M. D. Virto, L. C. Landeta, I. Agud, R. Solozabal, J. M. Lascaray, M. de Renobales, M. J. Llama, and L. Serra. 1993. "Immobilization of *Candida Rugosa* Lipase and Some Properties of the Immobilized Enzyme." *Enzyme and Microbial Technology* 15 (3):239–247.

Murty, V. R., J. Bhat, and P. K. A. Muniswaran. 2002. "Hydrolysis of Oils by Using Immobilized Lipase Enzyme: A Review." *Biotechnology and Bioprocess Engineering* 7 (2):57–66.

Nawani, N., R. Singh, and J. Kaur. 2006. "Immobilization and Stability Studies of a Lipase from *Thermophilic Bacillus* Sp: The Effect of Process Parameters on Immobilization of Enzyme." *Electronic Journal of Biotechnology* 9 (5):559–565.

Ogino, S. 1970. "Formation of the Fructose-Rich Polymer by Water-Insoluble Dextransucrease and Presence of Glycogen Value-Lowering Factor." *Agricultural and Biological Chemistry* 34:1268–1271.

Ozmen, E. Y., M. Sezgin, and M. Yilmaz. 2009. "Synthesis and Characterization of Cyclodextrin-Based Polymers as a Support for Immobilization of Candida Rugosa Lipase." *Journal of Molecular Catalysis B: Enzymatic* 57 (1–4):109–114.

Padmini, P., S. K. Rakshit, and A. Baradarajan. 1993. "Studies on Immobilization of Lipase on Alumina for Hydrolysis of Ricebran Oil." *Bioprocess and Biosystems Engineering* 9 (1):43–46.

Pahujani, S., S. S. Kanwar, G. Chauhan, and R. Gupta. 2008. "Glutaraldehyde Activation of Polymer Nylon-6 for Lipase Immobilization: Enzyme Characteristics and Stability." *Bioresource Technology* 99 (7):2566–2570.

Palomo, J., G. Fernández-Lorente, C. Mateo, M. Fuentes, R. Fernández-Lafuente, and J. M. Guisán. 2002a. "Modulation of the Enantioselectivity of *Candida antarctica B* Lipase Via Conformational Engineering. Kinetic Resolution of (±)-α-Hydroxy-Phenylacetic Acid Derivatives." *Tetrahedron: Asymmetry* 13 (12):1337–1345.

Palomo, J., G. Muñoz, G. Fernández-Lorente, C. Mateo, R. Fernández-Lafuente, and J, M. Guisán. 2002b. "Interfacial Adsorption of Lipases on Very Hydrophobic Support (Octadecyl-Sepabeads): Immobilization, Hyperactivation and Stabilization of the Open Form of Lipases." *Journal of Molecular Catalysis B: Enzymatic* 19–20:279–286.

Palomo, J., G. Muñoz, G. Fernández-Lorente, C. Mateo, M. Fuentes, J. M. Guisán, and R. Fernández-Lafuente. 2003. "Modulation of *Mucor Miehei* Lipase Properties Via Directed Immobilization on Different Hetero-Functional Epoxy Resins: Hydrolytic Resolution of (R,S)-2-Butyroyl-2-Phenylacetic Acid." *Journal of Molecular Catalysis B: Enzymatic* 21 (4–6):201–210.

Pronk, W., P. J. A. M. Kerkhof, C. van Helden, and K. van't Riet. 1988. "Hydrolysis of Triglycerides by Immobilized Lipase in a Hydrophilic Membrane Reactor." *Biotechnology and Bioengineering* 32 (4–5):512–518.

Rajendran, A., A. Palanisamy, and V. Thangavelu. 2009. "Lipase Catalyzed Ester Synthesis for Food Processing Industries." *Brazilian Archives of Biology and Technology* 52:207–219.

Ramani, K., R. Boopathy, A. B. Mandal, and G. Sekaran. 2011. "Preparation of Acidic Lipase Immobilized Surface-Modified Mesoporous Activated Carbon Catalyst and Thereof for the Hydrolysis of Lipids." *Catalysis Communications* 14 (1):82–88.

Reetz, M. T., A. Zonta, and J. Simpelkamp. 1996a. "Efficient Immobilization of Lipases by Entrapment in Hydrophobic Sol-Gel Materials." *Biotechnology and Bioengineering* 49 (5):527–534.

Reetz, M. T., A. Zonta, and J. Simpelkamp, and W. Konen. 1996b. "In Situ Fixation of Lipase-Containing Hydrophobic Sol-Gel Materials on Sintered Glass—Highly Efficient Heterogeneous Biocatalysts." *Chemical Communications* (11):1397–1398.

Reetz, M. T., R. Wenkel, and D. Avnir. 2000. "Entrapment of Lipase in Hydrophobic Sol-Gel-Materials: Efficient Heterogeneous Biocatalysts in Aqueous Medium." *Synthesis* 6:781–783.

Rodrigues, A. R., J. M. S. Cabral, and M. A. Taipa. 2002. "Immobilization of Chromobacterium Viscosum Lipase on Eudragit S-100: Coupling, Characterization and Kinetic Application in Organic and Biphasic Media." *Enzyme and Microbial Technology* 31 (1–2):133–141.

Roig, M. G., J. F. Bello, F. G. Velasco, C. D. de Celis, and J. M. Cachaza. 1987. "Biotechnology and Applied Biology Section Applications of Immobilized Enzymes." *Biochemical Education* 15 (4):198–208.

Rucka, M., and B. Turkiewicz. 1990. "Ultrafiltration Membranes as Carriers for Lipase Immobilization." *Enzyme and Microbial Technology* 12 (1):52–55.

Scherer, R., J. V. Oliveira, S. Pergher, and D. de Oliveira. 2011. "Screening of Supports for Immobilization of Commercial Porcine Pancreatic Lipase." *Materials Research* 14:483–492.

Shafei, M. S., and R. F. Allam. 2010. "Production and Immobilization of Partially Purified Lipase from Penicillium Chrysogenum." *Malaysian Journal of Microbiology* 6 (2):196–202.

Shamel, M. M., R. B. Azaha, and S. Al-Zuhair. 2005. "Adsorption of Lipase on Hollow Fiber Membrane Chips." *Artificial Cells, Blood Substitutes, and Immobilization Biotechnology* 3 (4):423–433.

Shaw, J.-F., R.-C. Chang, F. F. Wang, and Y. J. Wang. 1990. "Lipolytic Activities of a Lipase Immobilized on Six Selected Supporting Materials." *Biotechnology and Bioengineering* 35 (2):132–137.

Tahoun, M. K. 1986. "Production of Fatty Acids and Partial Glycerides from Milk Fat Triglycerides by Immobilized Candida Cylindrecea Lipase." In *MILK the Vital Force*, edited by International Dairy Federation, 219. Springer.

Taylor, F., C. C. Panzer, J. C. Craig Jr., and D. J. O'Brien. 1986. "Continuous Hydrolysis of Tallow with Immobilized Lipase in a Microporous Membrane." *Biotechnology and Bioengineering* 28 (9):1318–1322.

Tsai, S.-W., and S.-S. Shaw. 1998. "Selection of Hydrophobic Membranes in the Lipase-Catalyzed Hydrolysis of Olive Oil." *Journal of Membrane Science* 146 (1):1–8.

Ursoiu, A., C. Paul, C. Marcu, M. Ungurean, and F. Péter. 2011. "Double Immobilized Lipase for the Kinetic Resolution of Secondary Alcohols." *World Academy of Science, Engineering and Technology* 76:70–74.

Villeneuve, P., J. M. Muderhwa, J. Graille, and M. J. Haas. 2000. "Customizing Lipases for Biocatalysis: A Survey of Chemical, Physical and Molecular Biological Approaches." *Journal of Molecular Catalysis B: Enzymatic* 9 (4–6):113–148.

Walker, J. M., and R. Rapley. 2009. *Molecular Biology and Biotechnology*. Royal Society of Chemistry.

Wang, W., Y. Xu, X. Qin, D. Lan, B. Yang, and Y. Wang. 2014. "Immobilization of Lipase Smg1 and Its Application in Synthesis of Partial Glycerides." *European Journal of Lipid Science and Technology* 116 (8):1063–1069.

Watanabe, T., Y. Suzuki, Y. Sagesaka, and M. Kohashi. 1994. "Immobilization of Lipases on Polyethylene and Application to Perilla Oil Hydrolysis for Production of Alpha-Linolenic Acid." *Journal of Nutritional Science and Vitaminology* 41 (3):307–312.

Yadav, G. D., and S. R. Jadhav. 2005. "Synthesis of Reusable Lipases by Immobilization on Hexagonal Mesoporous Silica and Encapsulation in Calcium Alginate: Trans-esterification in Non-Aqueous Medium." *Microporous and Mesoporous Materials* 86 (1–3):215–222.

Yahya, A. R. M., W. A. Anderson, and M. Moo-Young. 1998. "Ester Synthesis in Lipase-Catalyzed Reactions." *Enzyme and Microbial Technology* 23 (7–8):438–450.

Yang, G., J. Wu, G. Xu, and L. Yang. 2009. "Improvement of Catalytic Properties of Lipase from Arthrobacter Sp. By Encapsulation in Hydrophobic Sol-Gel Materials." *Bioresource Technology* 100 (19):4311–4316.

Yilmaz, E., and M. Sezgin. 2012. "Enhancement of the Activity and Enantioselectivity of Lipase by Sol-Gel Encapsulation Immobilization onto *B*-Cyclodextrin-Based Polymer." *Applied Biochemistry and Biotechnology* 166 (8):1927–1940.

Yong, Y., Y.-X. Bai, Y.-F. Li, L. Lin, Y.-J. Cui, and C.-G. Xia. 2008. "Characterization of *Candida Rugosa* Lipase Immobilized onto Magnetic Microspheres with Hydrophilicity." *Process Biochemistry* 43 (11):1179–1185.

Yong, H. S., B. T. Tey, S. L. Hii, S. M. M. Kamal, A. Ariff, and T. C. Ling. 2010. "Application of a High Density Adsorbent in Expanded Bed Adsorption of Lipase from *Burkholderia Pseudomallei*." African *Journal of Biotechnology* 9 (2):203–216.

Zarcula, C., C. Kiss, L. Corici, R. Croitoru, C. Csunderlik, and F. Peter. 2009. "Combined Sol-Gel Entrapment and Adsorption Method to Obtain Solid-Phase Lipase Biocatalyts." *Revista de Chimie (Bucharest)* 60 (9):922–927.

4 Kinetics of Soluble and Immobilized Enzymes

Immobilized enzyme reactors are increasingly popular due to their advantages over conventional catalysts. For efficient reactor design and performance prediction, quantitative knowledge of reaction kinetics and the factors affecting them is required. In this chapter, enzyme catalytic mechanisms are described and the kinetic models developed from these mechanisms are discussed. The chapter also discusses the kinetics of immobilized enzymes and their related mass transfer effects. Diffusion restrictions are described with a particular focus on packed bed reactors. The chapter concludes with a brief discussion of immobilized enzyme reactor design and scale-up.

4.1 ENZYME ACTIVITY

The enzyme activity is used as an index to estimate the enzyme's potential to produce the desired product. Different physicochemical techniques are available to measure the activity, either by substrate consumption or product formation (Smeltzer et al., 1992). For a good estimate of the activity, the range of substrate concentrations should be carefully selected and accurately measured. Continuous assays, which continuously monitor changes in the reaction solution's physical properties, such as light absorbance (colorimetric), fluorescence (fluorometric), or heat release/absorbance (calorimetric), are among the possible techniques. Discontinuous assays, where samples from a reaction solution are collected at intervals, and the amount of substrate or product concentrations are measured, are also frequently used.

Quantification can be made by adding a reagent that does not interfere with the reaction and converts the unchanged substrates or products to a measurable compound, or by separating the desired product from the reaction mixture and measuring the concentration. Although continuous assays are more valuable, not all reaction products generate a readily measurable signal. To use these assays, auxiliary enzymes that produce measurable changes should be added.

Lipase activity can be measured using the aforementioned methods and are similar to other enzymatic reactions (Beisson et al., 2000). Choosing the appropriate method depends on lipase sensitivity, the availability of substrates, and the ease of measurement (Hendrickson, 1994; Singh and Mukhopadhyay, 2012).

4.2 REACTION MECHANISMS

Enzymatic reactions are usually initiated at a bond between the enzyme and the substrate. The active sites of the enzyme have similar dimensions to that of the substrate, which allows easy access and binding. When the substrate binds to active sites in the enzyme with the correct orientation, it reacts and the product is produced. Typically, the catalytic mechanism of the enzyme is a complex series of steps that specify the reaction kinetics and identify the affinity of the enzyme toward binding with reaction substrates. The mechanisms fall into two classes: one-substrate and multisubstrate mechanisms, depending on the number of substrates that appear in the reaction.

4.2.1 ONE-SUBSTRATE REACTION

The effect of substrate concentration on enzymatic reaction was first put forward in 1903 (Henri, 1903), where the conversion into the product involved a reaction between the enzyme and the substrate to form a substrate–enzyme complex that is then converted to the product. However, the reversibility of the substrate–enzyme complex and its final breakdown into the substrate and free enzyme regeneration was generally ignored. In 1913, Michaelis and Menten took this into consideration and proposed the scheme shown in Equation 4.1 for a one-substrate enzymatic reaction. Experimental data, that is, the initial reaction rates, were collected to support their analysis. The reaction mechanism, which is one of the most common mechanisms in enzymatic reactions, was based on the assumption that only a single substrate and product are involved in the reaction.

$$E + S \underset{k_{-1}}{\overset{k_1}{\rightleftharpoons}} ES \xrightarrow{k_2} P + E \tag{4.1}$$

where S represents the substrate; E the enzyme; ES the enzyme–substrate complex; P the product; k_1 and k_{-1} are ES formation and dissociation constants, respectively; and k_2 is the constant of product formation and release from the active site. This is usually referred to as the turnover number.

Although the reaction mechanism shown in Equation 4.1 is commonly used, a more representative mechanism should have some degree of reversibility in considering the enzyme–substrate complex's reversible conversion to the product, as shown in Equation 4.2:

$$E + S \underset{k_{-1}}{\overset{k_1}{\rightleftharpoons}} ES \underset{k_{-2}}{\overset{k_2}{\rightleftharpoons}} P + E \tag{4.2}$$

where k_{-2} is the product dissociation constant.

4.2.2 MULTISUBSTRATE REACTIONS

Multiple substrate reactions are more frequently occurred than single substrate reactions. Thus, efforts have been made to develop two or more substrate enzymatic reaction mechanisms. One of the most important examples is lipid hydrolysis, where lipid and water molecules act as two substrates to produce two products; namely fatty acids and glycerol. Cleland (1963) proposed three types of mechanism for two substrate reactions based on the order of adding the substrates and products release from the active site within the reaction sequence. These are ordered-sequential, random-sequential, and Ping-Pong, shown in Figure 4.1.

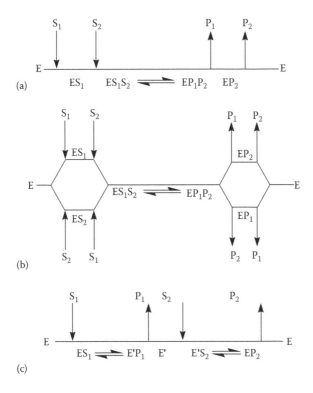

FIGURE 4.1 Graphical representation of (a) ordered-sequential, (b) random-sequential, and (c) Ping-Pong mechanisms.

The difference between the sequential reactions shown in Figure 4.1a,b and the Ping-Pong mechanism shown in Figure 4.1c is that in the sequential mechanisms, each substrate combines with the enzyme before the reaction occurs and forms a ternary complex, whereas in the Ping-Pong mechanism the first product is formed and released before the second substrate binds to the enzyme. This means that the enzyme in the Ping-Pong mechanism exists on two active sites where each recognizes one substrate and has two complexes present.

In the ordered-sequential mechanism, there is no formation of a second-substrate complex, as the enzyme initially has no site for binding with the second substrate. This site is formed only when the first substrate has been bound to the enzyme. On the other hand, in the random mechanism, the enzyme can bind with either the first or second substrate to produce the corresponding complex, which can then bind to the other substrate to generate the new complex, which later generates and releases the product.

4.3 KINETIC MODELS

Kinetic modeling is a quantitative analysis of every factor that determines the enzyme catalytic prospective and activity. It uses a maximum enzyme potential that can be determined by initial rate studies. The initial reaction rate, v_o, is the rate

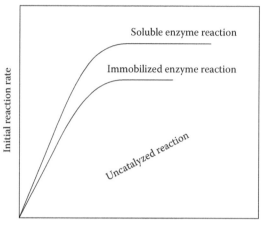

FIGURE 4.2   Effect of initial substrate concentrations on the initial reaction rate.

determined at, or near, zero time of the reaction, where the concentration of reaction substrates is high and does not change from its initial value. Most important, by using the initial rate concept, the effect of product inhibition and enzyme insolubility are eliminated.

The reaction rate increases with the increase in substrate concentration, as a result of having more substrate interactions with active sites. It forms enzyme–substrate complexes, as shown in Figure 4.2. These represent the effect of the initial substrate concentration on the initial reaction rate. However, at high substrate concentrations, the enzyme becomes saturated with substrates and there are only few free active sites left available. Therefore, increasing the substrate concentration any further does not make much difference to the reaction rate. Thus, at low substrate concentration, the reaction rate increases directly with the increase in substrate concentration. At higher concentrations, this relationship becomes curvilinear and the rate eventually reaches a maximum independently of substrate concentration. This is a distinctive property of binding mechanisms. To extend and prolong the linear relation, enzyme concentration should be reduced. When enzyme activity decreases, the maximum plateau value is also reduced. This occurs due to enzyme instability and deactivation over time.

In Michaelis and Menten's (1913) analysis, the equilibrium between the complex formation and its dissociation is assumed to exist. However, in some cases, this assumption does not hold, especially when the product formation rate from the complex (k_2) is close to the rate of complex dissociation to the enzyme and substrate (k_{-1}). In such cases, a more general assumption, known as quasi-steady state, is used.

4.3.1   Rapid Equilibrium Approach

The most direct approach to analyze enzyme kinetics is by using rapid equilibrium assumption. For the equilibrium to exist, the conversion of the complex compound into product should be much slower than the rate of its formation, and reaction rate is limited by product formation. Michaelis and Menten (1913) showed that the enzyme–substrate curve, shown in Figure 4.2, can be expressed mathematically by Equation 4.3, which is characterized by two parameters. They are (1) v_{max}, which is the maximum rate when the enzyme is fully saturated with substrate, and equal to $k_2[E]$, and (2) K_M, which is the Michaelis-Menten constant that is equal to $[S]$ when v is half the value of v_{max}. The latter indicates the affinity of the substrate toward the enzyme, where lower values indicate higher affinities. When the K_M value is much higher than the substrate concentration, the reaction rate will be sensitive to any changes in substrate concentration, and vice versa.

$$v = \frac{v_{max}[S]}{K_M + [S]} \tag{4.3}$$

Typically, the enzyme exists either as a free enzyme or in a complex with the substrate. The substrate exists as a free or in a complex with the enzyme. The conservation equations of the enzyme and substrate are given by Equations 4.4 and 4.5, respectively

$$[E]_o = [E] + [ES] \tag{4.4}$$

$$[S]_o = [S] + [ES] \tag{4.5}$$

where $[E]_o$ is the total amount of enzymes in the system that exists either in free form $[E]$ or in the complex $[ES]$. The binding constant (K_s) of irreversible product formation is given by Equation 4.6:

$$K_s = \frac{[E][S]_o}{[ES]} \tag{4.6}$$

As the rate of enzyme-catalyzed reactions is always proportional to product concentration, the initial (v_o) reaction rate can be expressed as in Equation 4.7. The subsequent steps of the derivation are illustrated in Equations 4.8 to 4.10:

$$\text{Rate equation: } v_o = k_2[ES] \tag{4.7}$$

$$\text{Derivation: } \frac{v_o}{[E]_o} = \frac{k_2[ES]}{[E] + [ES]} \tag{4.8}$$

$$= \frac{k_2}{\dfrac{[E]}{[ES]} + 1} \tag{4.9}$$

$$= \frac{k_2}{\dfrac{K_s}{[S]_o} + 1} \tag{4.10}$$

For a case when $[E]_o \ll [S]_o$, $[ES]$ would be much smaller than $[S]_o$, and $[S] = [S]_o$, and $v = v_o$, as shown in Equation 4.11:

$$v_o = v_{max} = k_2[E]_o \qquad (4.11)$$

On the other hand, at low $[S]_o$, Equation 4.3 is reduced to Equation 4.12, suggesting first-order dependence of v_o to $[E]_o$ and $[S]$, with linear dependence on $\dfrac{k_2}{K_M}[E]_o$:

$$v_o = \frac{v_{max}}{K_M}[S] = \frac{k_2}{K_M}[E]_o[S] \qquad (4.12)$$

4.3.2   QUASI-STEADY STATE

For reactions where product formation is a limiting step, the difference between product formation (k_2) and complex dissociation rates is small. With a steady-state assumption, the rate of the formation of substrate complex and the buildup concentration rate (k_1) is assumed to be equal to that of its dissociation (k_{-1}), suggesting that $[ES]$ does not change during the course of the reaction. Derivation of reaction rates based on a steady-state assumption is similar to that of an equilibrium assumption. The key difference is in the method of expressing $[ES]$. The steady-state condition is shown in Equation 4.13, and the derivations are given in Equations 4.14 to 4.19.

$$\text{Steady-state condition: } \frac{d[ES]}{dt} = 0 \qquad (4.13)$$

$$k_1[E][S] = (k_{-1} + k_2)[ES] \qquad (4.14)$$

$$[ES] = \frac{k_1}{(k_{-1} + k_2)}[E][S] \qquad (4.15)$$

$$\frac{v_o}{[E]_o} = \frac{k_2[ES]}{[E] + [ES]} \qquad (4.16)$$

$$= \frac{k_2\left\{\dfrac{k_1}{(k_{-1} + k_2)}[E][S]\right\}}{[E] + \left\{\dfrac{k_1}{(k_{-1} + k_2)}[E][S]\right\}} \qquad (4.17)$$

$$v_o = \frac{k_2[E]_o[S]}{\left(\dfrac{k_{-1} + k_2}{k_1}\right) + [S]}$$ (4.18)

$$v = \frac{v_{max}[S]}{K_m + [S]}, \text{ where } K_m = \frac{k_{-1} + k_2}{k_1}$$ (4.19)

When $k_2 = k_{-1}$, K_m can be simplified to $\dfrac{k_{-1}}{k_1}$, which is identical to K_M derived from an equilibrium assumption.

The derivation of the reversible rate equation is similar to that of the irreversible equation. However, the difference is in both the forward and reverse reaction rates for the product. See Equation 4.20. It can be converted to an irreversible reaction by considering [P] as equal to zero.

$$v_o = \frac{K_{M,P}\, v_{1,max}[S] - K_{M,S}\, v_{2,max}[P]}{K_{M,S} K_{M,P} + K_{M,P}[S] + K_{M,S}[P]}$$ (4.20)

The same approach can be used to derive multisubstrate reaction rates. For reactions involving more than two substrates, reaction rates are more complex than in the simpler Michaelis-Menten model (Equation 4.3) and the rate constants cannot be as easily derived. The number of constants defined is large and none of them are related to Michaelis-Menten constants. For example, the expression rate of a reaction containing three substrates and three products has 27 terms in the denominator. In two-substrate mechanisms, both order and random sequential models have the same expression rate (Bisswanger, 2008). Equations 4.21 and 4.22 show the reaction rates for sequential (order and random) and Ping-Pong models, respectively, where the main difference between the sequential (order and random) and Ping-Pong kinetic equations is the constant term present in the denominator:

$$v = \frac{v_{max}[S_1][S_2]}{K_{S_1} K_{S_1 S_2} + K_{S_1 S_2}[S_1] + K_{S_2 S_1}[S_2] + [S_1][S_2]}$$ (4.21)

$$v = \frac{v_{max}[S_1][S_2]}{K_{S_1 S_2}[S_1] + K_{S_2 S_1}[S_2] + [S_1][S_2]}$$ (4.22)

4.4 DETERMINATION OF KINETIC PARAMETERS

The model parameters K_M and v_{max} found in Equation 4.3 can be determined using several approaches: (1) directly from the Michaelis-Menten model, (2) using Lineweaver and Burk's method (Lineweaver and Burk, 1934), (3) from the Eadie-Hofstee approach (Eadie, 1942; Hofstee, 1959), or (4) from Hanes-Woolf (Haldane, 1957) plots to determine initial reaction rates at different substrate concentrations. The most common approach is an algebraic transformation, or linearization, of data that takes the reciprocal of the Michaelis-Menten equation. This was proposed by Lineweaver and Burk (1934) and is shown in Equation 4.23:

$$\frac{1}{v} = \frac{K_M}{v_{max}} \frac{1}{[S]} + \frac{1}{v_{max}} \qquad (4.23)$$

The plot of the reciprocal initial rate verses reciprocal substrate concentrations yields a straight line on a slope equal to $\dfrac{K_M}{v_{max}}$, and an intercept on the vertical axis that equals $\dfrac{1}{v_{max}}$. Arguments have been put forward regarding the large margin of error generated by rate reciprocals if there is a small error in the initial measurement. However, this method is commonly used due to the separation of dependent and independent variables.

To overcome the limitation of the Lineweaver-Burk method, another graphic method for K_M and v_{max} determination has been proposed, where linearization is achieved by the multiplication of both sides of Equation 4.23 by $[S]$, as shown in Equation 4.24. A plot of $\dfrac{[S]}{v}$ versus $[S]$ gives a straight line with an intercept of $\dfrac{K_M}{v_{max}}$ and a slope of $\dfrac{1}{v_{max}}$. Errors are more consistently distributed in this method compared to the Lineweaver-Burk method.

$$\frac{[S]}{v} = \frac{1}{v_{max}}[S] + \frac{K_M}{v_{max}} \qquad (4.24)$$

Another approach is to use a nonlinear regression technique, which produces a weighted least square that maximizes the efficiency of the parameter estimation. This technique does not require linearization and can be used to determine multiparameters. Hernandez and Ruiz (1998) developed an Excel template for the calculation of enzyme kinetic parameters using this technique.

The kinetic parameters in Equations 4.21 and 4.22 can be determined from experimental data using nonlinear regression techniques. Nevertheless, these equations can be simplified by considering the excess concentration of one of the substrates. For example, at high values of $[S_2]$, the reaction rate can be simplified to a Michaelis-Menten equation form.

4.5 FACTORS AFFECTING REACTION RATES

The rate of enzymatic reactions is affected by several factors. The most important are factors that disrupt the protein structure, such as reaction temperature and pH. Enzyme and inhibitor concentrations are the other factors that affect the reaction.

4.5.1   ENZYME LOADING

For excess substrate concentrations, based on the Michaelis-Menten assumption, the enzyme concentration increases as the rate of the reaction increases. This is due to more enzyme molecules (and thus more active sites) being available to catalyze the reaction. Therefore, more enzyme–substrate complexes are formed. However, in real situations a point is reached where every substrate molecule gets bound to the enzyme, and thus further increases in enzyme concentration do not increase the reaction rate. This point is usually referred to as substrate saturation. The Michaelis-Menten model does not take this into consideration.

4.5.2   TEMPERATURE

Since enzymes are proteins, any environmental factor that affects protein structure may change enzyme activity. Denaturation is the distraction of the secondary and tertiary structures of the enzyme. This can be reversible or irreversible. Denaturation is said to be reversible when the enzyme regains its native structure after removing the enlistment stick, whereas in irreversible denaturation, regaining its structure is not possible. Such structural changes are the result of bond disruption and changes in the ionization of the active sites.

For any chemical reaction, higher temperatures usually lead to higher reaction rates. Increasing the temperature increases collisions between molecules and results in sufficient energy to bypass the energy barrier and reach a transition state. This effect is best described using the Arrhenius equation shown in Equation 4.25. This suggests that the first-order rate constant is proportional to the exponential normalized activation energy (E_a) due to the gas constant (R).

$$k = A \exp\left(\frac{-E_a}{RT}\right) \tag{4.25}$$

where A is the Arrhenius constant, also known as the frequency factor. Although this is true for any chemical reaction, enzymatic reactions deviate from this relationship at high temperatures. Being a protein, an enzyme is denatured at high temperatures and its activity drops. Thus, there is an optimum reaction temperature at which the enzyme works effectively. If the temperature increases above this optimum value, which is usually close to the normal temperature of the organism where the enzyme was extracted, the activity of the enzyme declines sharply.

Each enzyme has its own optimum temperature for optimal activity. Enzymes that have been maintained longer at a temperature where the structure is stable have a higher chance of denaturation at higher temperatures. High thermal energy breaks down hydrogen bonds that hold the secondary and tertiary structure of the enzyme together. So, enzymes lose their shape and become random coils where the substrates no longer fit the active sites or bind properly. At the beginning of any reaction the active enzyme is equal to the total number of enzymes present in both free and complex forms. Generally, deactivation is first with respect to active enzyme concentration. The timeframe for enzyme deactivation is given in Equation 4.26:

$$\frac{d[E]_a}{dt} = -k_d[E]_a \qquad (4.26)$$

where k_d is the deactivation rate constant and $[E]_a$ is the active enzyme concentration. As mentioned earlier, the value of v_{max} depends on the number of active enzymes present. Therefore, as $[E]_a$ declines due to deactivation, so does the v_{max}. The deactivation constant, k_d, varies with temperature according to the Arrhenius correlation. Consequently, deactivation rates increase along with temperature, and hence the enzymatic reaction rate decreases. Equation 4.27 demonstrates the effect of temperature on the reaction rate:

$$v = \frac{A \, e^{-E_a/RT} E_o e^{-k_d t}[S]}{K_M + [S]} \qquad (4.27)$$

4.5.3   pH

Changing pH might alter the distribution of charge on the enzyme's active sites. Generally, denaturation of enzymes occurs at extremely low and high pH values. In these extreme conditions, ions interact with groups of enzyme amino acids and reduce the enzyme's ability to bind with substrates and the rate of the enzyme–substrate complex breakdown. The pH range at which an enzyme might be stable and active varies from one protein to another. However, most enzymes are stable at a pH value of 7.4 (Copeland, 2000; Covington and Covington, 2009) and commonly have bell-shaped curves with different pH values. Changes in pH alter the shape of enzymes as well as their active sites to a degree where the substrate cannot fit the active sites. Within a narrow range of pH values, enzyme deactivation is reversible. However, it is irreversible at extreme pH values. Thus, it is necessary to maintain reaction mixtures within the optimum pH range. To do so, reaction mixtures must be buffered by a component with a pKa value near to or at the same optimum pH as the enzyme. Equation 4.28 defines enzyme pH dependence for ionizing enzymes:

$$
\begin{array}{l}
E^- + H^+ \\
\quad \updownarrow \; K_2 \\
EH + S \;\underset{\longleftarrow}{\overset{K_M}{\longrightarrow}}\; EHS \;\xrightarrow{\;K_{cat}\;}\; EH + P \qquad (4.28) \\
+ \\
H \\
\quad \updownarrow \; K_1 \\
EH^{2+}
\end{array}
$$

The pH can also affect ionization of the substrate and reduce its concentration. The effect on ionized substrate is shown in Equation 4.29:

$$SH^+ + E \rightleftharpoons ESH^+ \longrightarrow E + HP^+$$

$$\Big\updownarrow K_1 \tag{4.29}$$

$$S + H^+$$

Equations 4.30 and 4.31 represent the rate equations for the two cases, respectively, where in both cases, Michael's constant, K_M, is affected.

$$v = \frac{v_{max}[S]}{K_M\left(1 + \dfrac{K_2}{[H^+]} + \dfrac{[H^+]}{K_I}\right) + [S]} \tag{4.30}$$

$$v = \frac{v_{max}[S]}{K_M\left(1 + \dfrac{K_1}{[H^+]}\right) + [S]} \tag{4.31}$$

4.5.4 INHIBITORS

There are many compounds that affect enzymatic reaction rates due to different mechanisms. Activators, such as cofactors and coenzymes, are compounds that bind with the enzyme and increase reaction rates. On the other hand, inhibitors are compounds that bind to the active site and reduce the rate by negatively influencing the catalytic properties of the enzyme's active sites (Panesar et al., 2010). In addition, an inhibitor can also bind at sites other than the enzyme's active site. For example, on the reverse, resulting in conformational changes in the active site and a decrease in catalytic activity.

This inhibitory effect can be reversible or irreversible. In the irreversible case the inhibitor, which is a poison in this instance, binds strongly to the enzyme, usually via covalent bonds. In the reversible case the inhibitor can be removed from the complex (Smith, 2010). Various forms of reversible inhibition have been investigated and classified based on the catalytic step at which they interact. Generally, when inhibitors bind to the enzyme, the $\dfrac{v_{max}}{K_M}$ magnitude decreases. If they bind to the enzyme complex, the v_{max} decreases.

4.5.4.1 Competitive Inhibition

Competitive inhibition occurs when the inhibitor, which has a similar structure to that of the substrate, binds to active sites on the enzyme in a manner that prevents substrate binding but does not result in forming a product. In this type of inhibition, the inhibitor is a compound that competes with the substrate for the same active site on the enzyme. The mechanism for such case is shown in Equation 4.32:

$$
\begin{array}{c}
\text{E} + \text{S} \underset{k_{-1}}{\overset{k_1}{\rightleftharpoons}} \text{ES} \xrightarrow{k_2} \text{E} + \text{P} \\
+ \\
\text{I} \\
\\
k_I \Big\Updownarrow k_{-I} \\
\\
\text{EI}
\end{array}
\tag{4.32}
$$

where I represents the inhibitor and EI is the enzyme–inhibitor complex.

In the presence of competitive inhibition, the reaction rate is dependent on both substrate and inhibitor concentrations. This type of inhibition increases the K_M value but does not affect the v_{max}, because the number of active sites is not altered. Due to inhibitor interference with the substrate, higher substrate concentrations are required to displace and outcompete the inhibitor at the binding site to reach the maximum rate. When the substrate concentration is sufficiently increased, the active sites will be occupied by the substrate and the inhibitor cannot bind. Thus, it would affect the $\frac{v_{max}}{K_M}$ value. The formation of an enzyme–inhibitor complex reduces the number of enzymes available to bind with the substrate, and as a result, the reaction rate decreases. Equation 4.33 shows a mathematical representation of a competitive inhibition rate:

$$
v = \frac{v_{max}[S]}{K_M \left(1 + \dfrac{[I]}{K_I}\right) + [S]}
\tag{4.33}
$$

4.5.4.2 Noncompetitive Inhibition

Noncompetitive inhibition occurs when the inhibitor binds to the enzyme at sites other than the active sites, as shown in Equation 4.34. This does not prevent substrate binding but results in [ESI] being inactive and so slows down the reaction rate. The extent of inhibition in this case depends on the inhibitor, not on the substrate concentration. This decreases the value of v_{max}, but does not interfere with K_M. Equation 4.35 shows the reaction rate expression.

$$
\begin{array}{c}
\text{E} + \text{S} \underset{k_{-1}}{\overset{k_1}{\rightleftharpoons}} \text{ES} \xrightarrow{k_2} \text{E} + \text{P} \\
+ \qquad\qquad\quad + \\
\text{I} \qquad\qquad\quad \text{I} \\
\\
\Big\Updownarrow KI_1 \qquad\qquad \Big\Updownarrow KI_2 \\
\\
\text{EI} + \text{S} \underset{k_{-5}}{\overset{k_5}{\rightleftharpoons}} \text{ESI}
\end{array}
\tag{4.34}
$$

$$v = \frac{v_{max}[S]}{K_M\left(1 + \dfrac{[I]}{K_{I_1}}\right) + \left(1 + \dfrac{[I]}{K_{I_2}}\right)[S]}$$ (4.35)

4.5.4.3 Uncompetitive Inhibition

Uncompetitive inhibition occurs when the inhibitor binds to the enzyme–substrate complex instead of the enzyme, resulting in a decrease v_{max} and K_M values. Equation 4.36 shows the mechanism and Equation 4.37 shows the reaction rate expression:

$$E + S \underset{k_{-1}}{\overset{k_1}{\rightleftharpoons}} ES \overset{k_2}{\longrightarrow} E + P$$
$$+$$
$$I$$
$$\Big\updownarrow K_1$$
$$ESI$$

(4.36)

$$v = \frac{v_{max}[S]}{K_M + \left(1 + \dfrac{[I]}{K_I}\right)[S]}$$ (4.37)

Determination of the inhibition kinetic parameter follows the same approach as simple Michaelis-Menten kinetics, discussed earlier, using the linearization approach. Thus, it is better to express Equations 4.33, 4.35, and 4.37 in terms of apparent parameters, as shown in Equation 4.38. This is compatible with a simple Michaelis-Menten relationship.

$$v = \frac{v_{max,app}[S]}{K_{M,app} + [S]}$$ (4.38)

Experiments are designed in a matrix where initial reaction rates are collected at different substrate and inhibitor concentrations. Linearization of the initial rate is given in Equation 4.39, and Table 4.1 represents the apparent constants for each of the three inhibition mechanisms. A graphic representation is shown in Figure 4.3. Once the mechanism has been identified the parameters can be identified.

$$\frac{1}{v} = \frac{K_{M,app}}{v_{max,app}}\frac{1}{[S]} + \frac{1}{v_{max,app}}$$ (4.39)

For a plug flow reactor, Equation 4.3 can be rewritten as

$$-\frac{d[S]}{dt} = \frac{v_{max}[S]}{K_M + [S]}$$ (4.40)

TABLE 4.1

Apparent Kinetic Parameter for Different Inhibition Mechanisms

Mechanism	$v_{max,app}$	$K_{M,app}$
Competitive inhibition	v_{max}	$K_M\left(1+\dfrac{[I]}{K_I}\right)$
Noncompetitive inhibition	$\dfrac{v_{max}}{\left(1+\dfrac{[I]}{K_{I2}}\right)}$	$K_M\dfrac{\left(1+\dfrac{[I]}{K_{I1}}\right)}{\left(1+\dfrac{[I]}{K_{I2}}\right)}$
Uncompetitive inhibition	$\dfrac{v_{max}}{\left(1+\dfrac{[I]}{K_I}\right)}$	$\dfrac{K_M}{\left(1+\dfrac{[I]}{K_I}\right)}$

where $-\dfrac{d[S]}{dt}$ represents the rate of substrate consumption. Using the boundary conditions shown in Equations 4.41 and 4.42 and rearranging and integrating Equation 4.40, the reaction time can be as determined using Equation 4.43:

$$\text{at } t = 0, \ [S] = [S_o] \tag{4.41}$$

$$\text{at } t > 0, \ [S] = [S]_t, \tag{4.42}$$

$$K_M \text{Ln}\left(\frac{[S_o]}{[S]_t}\right) + [S_o] - [S]_t = v_{max}t \tag{4.43}$$

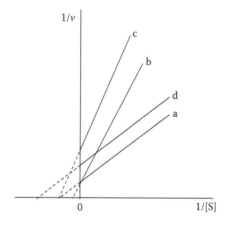

FIGURE 4.3 Graphical representation of Burke double reciprocal plots for (a) without inhibition, (b) competitive inhibition, (c) noncompetitive inhibition, and (d) uncompetitive inhibition.

4.6 KINETICS OF IMMOBILIZED ENZYMES

For industrial uses of enzymes, processes have to be technically and economically feasible. Immobilization of the enzyme, which is usually carried by entrapment within a porous solid matrix, has advantages over soluble enzymes (Pereira et al., 2001; Tramper et al., 2001). This includes ease of reuse and enhancement of both enzyme activity and selectivity (D'Souza, 2001; Gianfreda and Scarfi, 1991). Immobilized enzymes have applications in various industries and in different reactor configurations. Among these reactors, packed bed reactors have shown promise in many processes. This is widely used with an immobilized enzyme system due to long retention times and ease of operation (Abu-Reesh, 1997).

On the other hand, the use of immobilized enzymes in continuous operation may have a negative effect on reaction kinetics. This can be either by conformational (structural), steric, or microenvironmental change (Kalthod and Uckenstein, 1982). The main shortcoming lies in mass transfer limitations that affect the activity of the enzyme, especially if immobilized on the internal surface of a porous support.

4.6.1 DIFFUSIONAL RESTRICTIONS

When immobilized enzymes are used, a heterogeneous system with two reaction phases—bulk and solid enzyme—exists. The reaction takes place in the solid phase, whereas reaction substrates and products dissolve in the bulk phase. Unlike soluble enzymes, substrate concentrations that vary with reaction progress differ from those at the active site (Malcata et al., 1990) and this results in different reaction rates. The gradient in the substrate concentration is usually caused by interparticle (external) and intraparticle (internal) diffusions, as shown in Figure 4.4, resulting in an increase in the time required to reach a steady state or certain conversion. A prompt test can be carried out to assess the presence of mass transfer limitations by plotting the double reciprocal of the initial reaction data at different bulk substrate concentrations. A straight line indicates an insignificant mass transfer effect; however, when immobilized enzymes are used, a nonlinear characteristic is usually observed at low substrate concentrations. This could be explained by mass transfer resistance caused by immobilization that depends on substrate diffusion at the active site.

4.6.1.1 External Mass Transfer Effect

When a chemical reaction occurs on an active site, diffusion and chemical reaction rates are in a steady state and take place simultaneously. In a reactor packed with enzymes immobilized in a nonporous support in a solid–fluid system, where enzyme particles are in contact with flowing fluid, the velocity near the particle surface is low and a stagnant film surrounding the particle exists. In such cases, for the reaction to occur, the substrate must first transfer from the bulk to the surface by molecular diffusion (Geankoplis, 1978). The concentration gradient across the stagnant film surrounding the particle results in a decrease in the reaction rate. This was clearly evident in most studies using immobilized enzymes on nonporous pellets (Cooney, 1991; Murugesan and Sheeja, 2005; Nath and Chand, 1996).

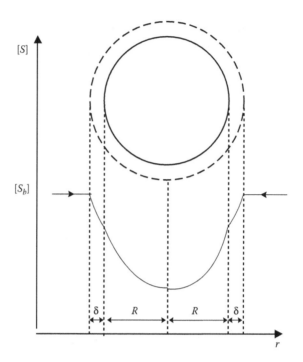

FIGURE 4.4 Concentration gradient of chemical reaction in an immobilized enzyme in a porous support.

The mass transfer rate from the bulk fluid to the surface of the immobilized enzyme is directly proportional to the external mass transfer coefficient, mass transfer area, and the substrate concentration difference between the bulk and the external surface of the enzyme, as given by Equation 4.44:

$$N = k_L([S_b] - [S])$$ (4.44)

where N is the external mass transfer rate; $[S]$ and $[S_b]$ are the substrate concentrations at the surface and in the bulk fluid, respectively; and k_L is the mass transfer coefficient. The mass transfer coefficient, k_L, can be determined from a dimensionless correlation (Equation 4.45) that relates the Sherwood (Sh), Reynolds (Re), and Schmidt (Sc) numbers. The dimensionless parameters are defined in Equations 4.46 and 4.47.

$$Sh = \frac{k_L d_p}{D} = C_0 Re^{C_1} Sc^{C_2}$$ (4.45)

$$Re = \frac{\rho u_f d_p}{\mu}$$ (4.46)

$$Sc = \frac{\mu}{\rho D}$$ (4.47)

where C_O, C_1, and C_2 are the adjustable parameters; d_p is the diameter of the catalyst particle; D is the molecular diffusion of the substrate; ρ and μ are the density and viscosity of the fluid, respectively; and u_f is the superficial velocity of the fluid. The mass transfer coefficient can also be determined from the substrate diffusion, D, film thickness, δ, and total surface area, A, as given in Equation 4.48:

$$k_L = \frac{D}{\delta} A \tag{4.48}$$

This mass transfer resistance and the reaction kinetics are usually compared using a dimensionless ratio, Damkohler number, Da, as in Equation 4.49. If Da is greater than unity, then the mass transfer resistance is significant and the reaction of the external diffusion is limited. On the other hand, when Da is less than unity, the reaction, in this case, is surface reaction limited.

$$Da = \frac{\text{reaction rate}}{\text{mass transfer rate}} = \frac{v_{max}/K_M}{k_L} \tag{4.49}$$

The external mass transfer resistance depends heavily on fluid flow conditions, such as temperature, pressure, and superficial velocity, in the reactor and the particle size of the catalyst. Varying these parameters can help to reduce the external diffusion restriction, for example, by increasing the velocity of the fluid phase over the particles.

4.6.1.2 Internal Mass Transfer

To increase the amount of immobilized enzyme per unit weight of the support, immobilization is carried out in porous supports with large internal surface areas. However, the substrate has to diffuse through the internal pores of the particle in order to reach the active site. This results in additional diffusion resistance. The substrate concentration profile in this case is shown in Figure 4.4. Internal diffusion resistance depends on particle size and shape, external substrate concentration, and effective diffusion within the particle, D_e. The latter depends on the molecular diffusion of the substrate in the support matrix, and on the porousness (ε_p) and tortuosity (τ) of the pores in the particle (Park et al., 2006; Young and Al-Duri, 1996). The effective diffusion can be determined from different mathematical expressions as shown in (Equations 4.50 to 4.52) (Pilkington et al., 1998):

$$D_e = \frac{\varepsilon_p^{3/2}}{D} \tag{4.50}$$

$$D_e = \varepsilon_p^2 D \tag{4.51}$$

$$D_e = \frac{\varepsilon_p D}{\tau} \tag{4.52}$$

where τ is the tortuosity as a function of the size and shape of the pores and the diffusing molecule, which generally has a value between 2 and 10 (Marrazzo et al., 1975).

Unlike external mass transfer, in internal mass transfer, the enzyme molecules are exposed to different environments according to their position on the matrix. This results in different concentration profiles that vary with particle geometry. Having said that, for simplicity, global determinations have to be considered. For ideal spherical enzyme particles, only radial derivatives are considered in the diffusion flux. Therefore, the differential mass balance with boundary conditions in the substrate, derived from Michaelis-Menten kinetics under steady-state conditions, with a negligible external diffusion effect can be derived as shown in Equation 4.53. The first condition (Equation 4.54) is only valid for a catalyst with negligible external diffusion limitations in respect to internal diffusion, whereas the second condition (Equation 4.55) reflects the fact that there is no substrate flux through the spherical particle center.

$$\frac{d^2[S]}{dr^2} + \frac{2}{r}\frac{d[S]}{dr} = v_{max}\frac{[S]}{D_e(K_M+[S])} \tag{4.53}$$

$$\text{at } r = R, [S] = [S_b] \tag{4.54}$$

$$\text{at } r = 0, \frac{d[S]}{dr} = 0 \tag{4.55}$$

Internal diffusion is usually characterized by a Thiele modulus, φ, defined in Equation 4.56. It is a dimensionless parameter that represents the relative effect of the reaction on mass transfer rate. It depends on kinetic properties and on mass transfer properties, such as effective diffusion of the substrate, catalyst particle size, and geometry.

$$\varphi = \hat{R}\sqrt{\frac{v_{max}}{K_M D_e}} \tag{4.56}$$

where $\hat{R}$ is the diffusion characteristic length in the catalyst particle and defined as the volume to surface area ratio. It equals R/3 for spherical particles, R/3 for cylindrical particles, and L for slabs. High values of φ indicate that either substrate diffusion is low or diffusion length is large enough to render the system mass transfer limited.

The number of these parameters is usually reduced by presenting the equations in a dimensionless form. For that, the following dimensionless parameters are usually defined. The normalized form of Equation 4.53 can be derived as given in Equation 4.57:

$$\frac{D^2[\bar{S}]}{d\xi^2} + \frac{2}{\xi}\frac{d[\bar{S}]}{d\xi} = 9\varphi^2\left(\frac{[\bar{S}]}{1+\beta[\bar{S}]}\right) \tag{4.57}$$

where

$$[\bar{S}] = \frac{[S]}{[S_b]} \quad \text{(dimensionless concentration)} \tag{4.58}$$

$$\xi = \frac{r}{R} \quad \text{(dimensionless radial distance)} \tag{4.59}$$

$$\beta = \frac{[S]}{K_M} \tag{4.60}$$

Associated boundary conditions for the preceding equation are

$$\text{at } \xi = 1: [\bar{S}] = 1 \tag{4.61}$$

$$\text{at } \xi = 0: \frac{d[\bar{S}]}{d\xi} = 0 \tag{4.62}$$

The concentration profile throughout the particle can be attained numerically by integrating Equation 4.57. The extent of mass transfer is commonly expressed by a local effectiveness factor, η, defined as the ratio between local reaction rates inside the catalyst particles and the reaction rate at the surface with bulk substrate concentration, as given by Equation 4.63. This can be rearranged to give Equation 4.64. The mean, or overall, effectiveness factor of the enzyme particles can be evaluated by an average integration of local effectiveness distribution inside the particle.

$$\eta = \frac{v}{\left(\dfrac{v_{max}[S_b]}{K_M + [S_b]} \right)} \tag{4.63}$$

$$\frac{[S_b]}{v} = \frac{1}{\eta} \left(\frac{[S_b]}{v_{max}} + \frac{K_M}{v_{max}} \right) \tag{4.64}$$

Differential equations are usually solved numerically by finite differences using the Crank-Nicolson method that can be implemented using any modeling software (Aseris et al., 2013; Curcio et al., 2006; Valencia et al., 2011) and orthogonal collocation that can transfer the problem to a system of algebraic equations (Chen and Chang, 1984; Long et al., 2003; Shiraishi et al., 1996). It can be also solved using a Runge-Kutta integration method (Huang and Chen, 1992; Lortie, 1994; Shiraishi et al., 1996; Vos et al., 1990). For simple kinetics, such as the Michaelis-Menten model, numerical values can be presented in a graphic plot that shows the relationship between effectiveness factors, the Thiele modulus, and a dimensionless substrate

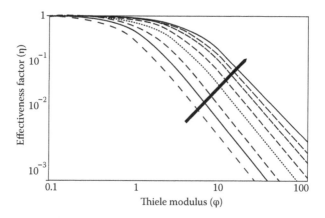

FIGURE 4.5 Effect of the Thiele modulus (φ) and substrate concentration on lactose hydrolysis. (Redrawn from Al-Muftah, A. E, and I. M. Abu-Reesh, 2005, *Biochemical Engineering Journal* 23:139–153. With permission.)

concentration. As in Figure 4.5, showing lactose hydrolysis by lactase, the effectiveness depends on the Thiele modulus value. A deeper understanding of effectiveness factors is given by Kasche and Schwegler (1979), where a correlation between calculated and experimental data was developed.

Generally, internal diffusion restrictions can be minimized and η approaches unity as φ decreases, or the substrate concentration increases. Therefore, for better utilization of the enzyme catalytic potential, the reactor should operate at a lower φ value, which can be manipulated by varying the particle size and the amount of immobilized enzyme per unit mass of the support, as shown in Figure 4.5.

To reduce internal mass transfer resistance, the use of small particles is advantageous. However, as the size of the catalysts decreases, their recovery becomes more complicated and costly, and can result in high drops in pressure. On the other hand, two effects are normally encountered with an increased enzyme loading per weight of immobilized catalyst. Although it increases the reaction rate, increasing the enzyme loading also increases internal mass transfer resistance. Therefore, for enzymes with a high specific activity, it is usually recommended to use immobilized enzymes with low loading.

Although, internal diffusion limitations restrict and reduce process efficiencies, in some cases they are advantageous, especially when substrate inhibition is present. In such cases, internal diffusion reduces the substrate concentration and inhibitor accumulation inside the catalyst particle. The inhibitory effect is then diminished resulting in an η value greater than unity.

4.6.2 MATHEMATICAL MODELING OF AN IMMOBILIZED BACKED REACTOR

Understanding mass transfer restrictions is important in evaluating the kinetic properties of the immobilized enzyme to achieve efficient bioreactor design. To develop a mathematical model that expresses the mass transfer effect, the following assumptions are usually made:

- The enzyme is distributed evenly within the particle.
- The reaction is isothermal and no pH change is involved.
- Constant diffusion coefficients.
- Spherical catalytic particles.
- Electrostatic effects are negligible.

Based on these assumptions, the following steady-state diffusion–reaction equations for mass balances of substrate concentrations within an immobilized enzyme can be written as

$$\frac{1}{r}\frac{d}{dr}\left(r\frac{d[S]}{dr}\right)+\frac{d^2[S]}{dz^2}=v_{max}\frac{[S]}{D_e(K_M+[S])} \tag{4.65}$$

where z represents the cylindrical coordinate. The initial and boundary conditions are shown in Equations 4.66 and 4.67:

$$\text{at } z=0,\ \frac{d[S]}{dz}=0 \tag{4.66}$$

$$\text{at } r=0,\ \frac{d[S]}{dr}=0 \tag{4.67}$$

For simplification, similar dimensionless parameters to those shown in Equation 4.57 are usually introduced, and Equation 4.65 can be rewritten in a normalized form with dimensionless parameters as in Equation 4.68:

$$\frac{1}{\xi}\frac{d}{d\xi}\left(\xi\frac{d[S]}{d\xi}\right)+\frac{d^2[S]}{dZ^2}=\frac{9\varphi^2[S]}{1+\beta[S]} \tag{4.68}$$

where

$$Z=\frac{z}{R}\ \text{(dimensionless cylindrical coordinate)} \tag{4.69}$$

4.7 IMMOBILIZED ENZYME BIOREACTOR DESIGN AND SCALE-UP

As noted earlier, enzymatic reactions rates not only depend on enzyme kinetics but also on other factors such as immobilization and mass transfer limitations. In addition, the amount of enzymes immobilized per weight of the support affects the overall cost of the process. High enzyme loading on low cost supports can be acceptable, especially if the enzyme cost is not high. Therefore, in the design of immobilized enzyme reactors, both enzyme characteristics and reactor design parameters should be considered.

For any chemical process, scale-up involves the transfer of a lab-scale process to the production scale so the desired product can be produced in large quantities. In many cases, two different scale systems perform differently in the same conditions. This is due to different flow patterns and residence time distribution. Unlike other chemical processes, the scale-up of a chemical reactor is a complex and difficult process, where chemical similarity (conversion and selectivity) depends on the interaction of mass transfer, kinetics, and hydrodynamics.

Dimensionless analysis is the most common engineering method used, where a set of carefully chosen dimensionless numbers, composed of combinations of physical properties and dimensions, are kept constant and equal. With the employment of scaling-up criteria, similar behavior in the different scales has to be validated. To predict the performance of the large-scale process, the geometrical, kinematic, thermal, hydrodynamic, and chemical similarities must be ensured. Geometric similarity is commonly used and can be achieved by fixing the dimensional ratios of the two reactors. For example, ratios of vessel height to diameter, stirrer diameter to vessel diameter, baffle width to vessel diameter, and stirrer clearance from the bottom of the vessel to the vessel diameter must be fixed when stirred tank reactors are scaled up. Kinematic similarity requires that flow velocities, usually represented in terms of Re number on the two different scales, have the same value. A similarity in kinetics is also important in processes where reaction mixtures involve immiscible fluids and where the residence time distribution has to be fixed. Furthermore, especially in biochemical reaction systems, the physiological similarities should be considered in order to get the same catalytic potential in the large-scale bioreactor as in the small one (Votruba and Sobotka, 1992). In such situations, the microenvironment has to be maintained.

In immobilized bed reactors, the scale-up is also dependent on parameters related to mass transfer as mass transfer resistance is expected to increase with scale. Therefore, a similarity in the mass transfer coefficient, which is usually affected by the changes in flow rate, reactor diameter, and fluid properties, is important in these processes.

REFERENCES

Abu-Reesh, I. M. 1997. "Predicting the Performance of Immobilized Enzyme Reactors Using Reversible Michaelis-Menten Kinetics." *Bioprocess Engineering* 17:131–137.

Al-Muftah, A. E., and I. M. Abu-Reesh. 2005. "Effects of Internal Mass Transfer and Product Inhibition on a Simulated Immobilized Enzyme-Catalyzed Reactor for Lactose Hydrolysis." *Biochemical Engineering Journal* 23:139–153.

Aseris, V., R. Baronas, and J. Kulys. 2013. "Computational Modeling of Bienzyme Biosensor with Different Initial and Boundary Conditions." *Informatica* 24 (4):505–521.

Beisson, F., V. Arondel, and R. Verger. 2000. "Assaying Arabidopsis Lipase Activity." *Biochemical Society Transactions* 28:773–775.

Bisswanger, H. 2008. *Enzyme Kinetics: Principles and Methods.* Wiley-VCH.

Chen, K.-C., and C.-M. Chang. 1984. "Operational Stability of Immobilized D-Glucose Isomerase in a Continuous Feed Stirred Tank Reactor." *Enzyme and Microbial Technology* 6 (8):359–364.

Cleland, W. W. 1963. "The Kinetics of Enzyme-Catalyzed Reactions with Two or More Substrates or Products: I. Nomenclature and Rate Equations." *Biochimica et Biophysica Acta* 67:104–137.

Cooney, D. O. 1991. "Determining External Film Mass Transfer Coefficients for Adsorption Columns." *AIChE Journal* 37 (8):1270–1274.

Copeland, R. A. 2000. *Enzymes: A Practical Introduction to Structure, Mechanism, and Data Analysis.* Wiley.

Covington, A., and T. Covington. 2009. *Tanning Chemistry: The Science of Leather.* Royal Society of Chemistry.

Curcio, S., V. Calabrò, and G. Iorio. 2006. "A Theoretical and Experimental Analysis of a Membrane Bioreactor Performance in Recycle Configuration." *Journal of Membrane Science* 273 (1–2):129–142.

D'Souza, S. F. 2001. "Immobilization and Stabilization of Biomaterials for Biosensor Applications." *Applied Biochemistry and Biotechnology* 96 (1–3):225–238.

Eadie, G. S. 1942. "The Inhibition of Cholinesterase by Physostigmine and Prostigmine." *Journal of Biological Chemistry* 146:85–93.

Geankoplis, C. J. 1978. *Transport Processes and Unit Operations.* Allyn & Bacon.

Gianfreda, L., and M. Scarfi. 1991. "Enzyme Stabilization: State of the Art." *Molecular and Cellular Biochemistry* 100 (2):97–128.

Haldane, J. B. S. 1957. "Graphical Methods in Enzyme Chemistry." *Nature* 179:932.

Hendrickson, H. S. 1994. "Fluorescence-Based Assays of Lipases, Phospholipases, and Other Lipolytic Enzymes." *Analytical Biochemistry* 219 (1):1–8.

Henri, V. 1903. *Lois Générales De L'action Des Diastases.* Librairie Scientifique A. Hermann.

Hernandez, A., and M. T. Ruiz. 1998. "An Excel Template for Calculation of Enzyme Kinetic Parameters by Non-Linear Regression." *Bioinformatics* 14 (2):227–228.

Hofstee, B. H. J. 1959. "Non-Inverted Versus Inverted Plots in Enzyme Kinetics." *Nature* 184:1296–1298.

Huang, T.-C., and D.-H. Chen. 1992. "Kinetic Studies on Urea Hydrolysis by Immobilized Urease in a Batch Squeezer and Flow Reactor." *Biotechnology and Bioengineering* 40 (10):1203–1209.

Kalthod, D. G., and E. Uckenstein. 1982. "Immobilized Enzymes: Electrokinetic Effects on Reaction Rates under External Diffusion." *Biotechnology and Bioengineering* 24 (10):2189–2213.

Kasche, K., and H. Schwegler. 1979. "Operational Effectiveness Factors of Immobilized Enzyme Systems." *Enzyme and Microbial Technology* 1:41–46.

Lineweaver, H., and D. Burk. 1934. "The Determination of Enzyme Dissociation Constants." *Journal of the American Chemical Society* 56 (3):658–666.

Long, W. S., S. Bhatia, and A. Kamaruddin. 2003. "Modeling and Simulation of Enzymatic Membrane Reactor for Kinetic Resolution of Ibuprofen Ester." *Journal of Membrane Science* 219 (1–2):69–88.

Lortie, R. 1994. "Evaluation of the Performance of Immobilized Enzyme Reactors with Michaelis-Menten Kinetics." *Journal of Chemical Technology & Biotechnology* 60 (2):189–193.

Malcata, F. X., H. Reyes, H. Garcia, C. Hill, and C. Amundson. 1990. "Immobilized Lipase Reactors for Modification of Fats and Oils—A Review." *Journal of the American Oil Chemists' Society* 67 (12):890–910.

Marrazzo, W. N., R. L. Merson, and B. J. McCoy. 1975. "Enzyme Immobilized in a Packed Bed Reactor: Kinetic Parameters and Mass Transfer Effects." *Biotechnology and Bioengineering* 17 (10):1515–1528.

Michaelis, L., and M. L. Menten. 1913. "The Kinetics of Invertase Action." *Biochemische Zeitschrift* 49:333–369.

Murugesan, T., and R. Y. Sheeja. 2005. "A Correlation for the Mass Transfer Coefficients During the Biodegradation of Phenolic Effluents in a Packed Bed Reactor." *Separation and Purification Technology* 42 (2):103–110.

Nath, S., and S. Chand. 1996. "Mass Transfer and Biochemical Reaction in Immobilized Cell Packed Bed Reactors: Correlation of Experiment with Theory." *Journal of Chemical Technology & Biotechnology* 66 (3):286–292.

Panesar, P. S., S. S. Marwaha, and H. K. Chopra. 2010. *Enzymes in Food Processing: Fundamentals and Potential Applications.* I. K. International.

Park, E. Y., M. Sato, and S. Kojumo. 2006. "Fatty Acid Methyl Ester Production Using Lipase-Immobilizing Silica Particles with Different Particle Sizes and Different Specific Surface Areas." *Enzyme and Microbial Technology* 39:889–896.

Pereira, F. B., H. F. De Castro, F. F. De Moraes, and G. M. Zanin. 2001. "Kinetic Studies of Lipase from *Candida Rugosa.*" In *Twenty-Second Symposium on Biotechnology for Fuels and Chemicals,* edited by B. H. Davison, J. McMillan, and M. Finkelstein, 739–752. Humana Press.

Pilkington, P. H., A. Margaritis, and N. A. Mensour. 1998. "Mass Transfer Characteristic of Immobilized Cells Used in Fermentation Processes." *Critical Reviews in Biotechnology* 18:237–355.

Shiraishi, F., T. Hasegawa, S. Kasai, N. Makishita, and H. Miyakawa. 1996. "Characteristics of Apparent Kinetic Parameters in a Packed-Bed Immobilized Enzyme Reactor." *Chemical Engineering Science* 51 (11):2847–2852.

Singh, A. K., and M. Mukhopadhyay. 2012. "Overview of Fungal Lipase: A Review." *Applied Biochemistry and Biotechnology* 166 (2):486–520.

Smeltzer, M. S., M. E. Hart, and J. J. Iandolo. 1992. "Quantitative Spectrophotometric Assay for Staphylococcal Lipase." *Applied and Environmental Microbiology* 58 (9):2815–2819.

Smith, D. M. 2010. "Protein Separation and Characterization Procedure." In *Food Analysis,* edited by S. S. Nielsen, 261–281. Springer.

Tramper, J., H. H. Beeftink, A. E. M. Janssen, L. P. Ooijkaas, J. L. van Roon, M. Strubel, and C. G. P. H. Schroën. 2001. "Biocatalytic Production of Semi-Synthetic Cephalosporins: Process Technology and Integration." In *Synthesis of Beta-Lactam Antibiotics: Chemistry, Biocatalysis and Process Integration,* edited by A. Bruggink, 206–249. Springer-Verlag.

Valencia, P., S. Flores, L. Wilson, and A. Illanes. 2011. "Effect of Internal Diffusional Restrictions on the Hydrolysis of Penicillin G: Reactor Performance and Specific Productivity of 6-Apa with Immobilized Penicillin Acylase." *Applied Biochemistry and Biotechnology* 165 (2):426–441.

Vos, H. J., P. J. Heederik, J. J. M. Potters, and K. Ch. A. M. Luyben. 1990. "Effectiveness Factor for Spherical Biofilm Catalysts." *Bioprocess Engineering* 5 (2):63–72.

Votruba, J., and M. Sobotka. 1992. "Physiological Similarity and Bioreactor Scale-Up." *Folia Microbiologica* 37 (5):331–345.

Young, Y. P., and B. Al-Duri. 1996. "Kinetic Studies on Immobilised Lipase Esterification of Oleic Acid and Octanol." *Journal of Chemical Technology & Biotechnology* 65:239–248.

5 Lipase-Catalyzed Reactions in Nonaqueous Media

Reacting lipophilic substrates with hydrophilic compounds, as in the case of most transesterification reactions, is one of the major difficulties in lipase-catalyzed reactions. Several parameters need to be considered to overcome this immiscibility problem. One commonly proposed strategy is the use of a nonaqueous medium. In this chapter, the advantages of using nonaqueous media in biochemical synthesis reactions, over aqueous and solvent-free systems, are discussed. The use of hydrophobic solvents is also discussed, followed by a presentation of the alternatives that can overcome the limitations of solvents. The focus of this chapter is mainly on the use of supercritical fluids (SCFs) as a green alternative reaction medium. The chapter also discusses ionic liquids (ILs) as another alternative. These solvents and the factors affecting their physical properties and their effect on the activity and stability of lipase are also discussed.

5.1 ORGANIC SOLVENTS

From a technical-economical point of view, carrying out lipase-catalyzed reactions in solvent-free systems is the most effective strategy. However, in some cases, the employment of organic solvents is also important. Conducting lipolytic reactions in an organic solvent media was suggested by Klibanov (1995). It has many advantages compared to solvent-free systems, such as better stability, ease of recovery, and the reuse of the immobilized enzyme from the reaction mixture. Generally, introducing an organic solvent into a synthetic reaction increases the solubility of substrates with similar polarity, thus increasing the reaction rate (Radzi et al., 2005). Moreover, conducting reactions in organic solvent media has other advantages, such as reducing the viscosity of the reaction medium and allowing a higher diffusion rate of substrates to the enzyme's active sites with lower mass transfer limitations (Klibanov, 2001; Zaks and Klibanov, 1984). On the other hand, lipase can only be active when the essential water molecules are present, and if those water molecules are completely stripped off, the integrity of the three-dimensional structure of the enzyme molecule, and hence its activity, are negatively affected. Although for most lipases actions a certain amount of water is essential in initiating the reaction, the control of the water quantity is crucial. In transesterification, excess water may result in undesired reactions, such as triglycerides hydrolysis (Chowdary and Prapulla, 2002; Leitner and Jessop, 2014; Macrae, 1983; Miller et al., 1991). Carrying out transesterification reactions in a solvent-free medium has other drawbacks besides mass transesterifications.

Short alcohols, which are the other substrates, have poor solubility in triglycerides, and when they are present as insoluble droplets in the organic phase, they strip the essential water molecules from the enzyme, resulting in deactivation (Shimada et al., 1999). This is further discussed in Chapter 6.

Although numerous organic solvents can be used, several aspects have to be considered when choosing a suitable one. These include; substrate polarity, solvent compatibility, inertness, low density, toxicity, and flammability (Adamczak and Krishna, 2004). The selection is critical, as some solvents may also strip the essential hydration shell from the lipase. The properties of the solvent, such as hydrophobicity (log P), which is the logarithm of an n-octanol-water partition coefficient, play a vital role in the selection (Khmelnitsky et al., 1991; Laane et al., 1987). Table 5.1 shows a list of common solvents used in enzymatically catalyzed reactions and their respective log P values. The selection of the proper organic solvent depends mainly on its application. For example, in triglyceride transesterification reactions, hydrophobic solvents typically result in better activity than hydrophilic solvents, due to the latter stripping off essential water molecules needed to keep the enzyme active (Cernia et al., 1998). This was clearly observed in the work done on the transesterification of

TABLE 5.1

Log P Values for Common Organic Solvents and SC-CO_2 Used in Enzyme-Catalyzed Reactions

Solvent	Log P	Solvent	Log P
iso-Octane[a]	4.7	CO_2[b,g]	0.9
n-Heptane[a]	4	*tert*-Butanol[a]	0.83
Petroleum ether[a]	3.8	Pyridine[a]	0.71
n-Hexane[a]	3.5	Aceton[a]	−0.26
Cyclohexane[a]	3.2	Acetonitrile[a]	−0.36
Toluene[a]	2.5	Ethylemethyle keton[a]	−0.8
CO_2[b,f]	2	Dioxane[a]	−1.1
Benzene[a]	2	[bmim][PF_6][c,d]	−2.39
Chloroform[a]	2	[bmim][NO_3][c,e]	−2.9

Source: Kaar, J. L. et al., 2003, *Journal of the American Chemical Society* 125 (14):4125–4131; Nakaya, H. et al., 2001, *Enzyme and Microbial Technology* 28 (2–3):176–182; Nie, K. et al., 2006, *Journal of Molecular Catalysis B: Enzymatic* 43 (1–4):142–147; Soumanou, M. M., and U. T. Bornscheuer, 2003, *Enzyme and Microbial Technology* 33 (1):97–103. With permission.

[a] Volatile organic.
[b] Supercritical fluid.
[c] Ionic liquid.
[d] 1-butyl-3-methylimidazolium hexafluorophosphate.
[e] 1-butyl-3-methylimidazolium nitrate.
[f] At 50°C and 118 bar.
[g] At 50°C and 3 bar.

methacrylate with 2-ethylhexanol using *Candida rugosa* lipase in different organic solvents. The lipase showed a higher activity in the hydrophobic solvent, *n*-hexane, compared to the hydrophilic solvent, butyl ether (Kamat et al., 1992). *n*-Hexane and *tert*-butanol are also widely used in oil and fat enzymatic reactions (Eltaweel et al., 2005; Hernández-Rodríguez et al., 2009; Li et al., 2006). However, they are both not as appropriate in fatty acid sugar esters, where the reaction substrates (sugar and fatty acids) differ in their polarity (Degn and Zimmermann, 2001; Sabeder et al., 2006; Tsitsimpikou et al., 1997) as sugars are polar, fatty acids are nonpolar, and the product (sugar esters) is amphiphilic. In such cases, the choice of solvent is more difficult.

Typically, the absence of water affects lipase conformation, as a small hydration layer is usually required for the interfacial activation of lipase (Dordick, 1989). The amount of water present or added to a reaction system affects the catalytic properties of the enzymes. In the beginning, the research focus was on reactions in media with a fixed total water content. However, Zaks and Klibanov (1988) verified that it is the water bound to the enzyme that determines the catalytic activity, not the total water content in the system. This is because water is equilibrated between the enzyme and the reaction solvent (Parker et al., 1995), rather than the total water content of the system. If the same quantity of water is added to systems containing organic solvents, the amount of water associated with the enzyme varies. More water is contained in more polar solvents. This concept allows for the quantity of water required in the reaction mixture to be expressed in terms of thermodynamic water activity (a_w) instead of water content, which depends on the type of lipase and solvent used. If reactions are carried out at a fixed a_w, then the amount of water associated with the lipase is fixed and this simplifies the prediction of lipase activity (Halling, 1994). Halling (1994) noted that when enzyme activity was compared among different media with the same water content, the results were confusing. However, when the water activity was fixed, the solvent's effect was flawless and predictable. For example, the amount of water required to obtain maximal activity in α-chymotrypsin and subtilisin (proteases) was much less in hydrophobic solvents when compared to hydrophilic solvents. This is due to the stripping of essential water from the enzyme, with hydrophilic solvents, which leads to enzyme denaturation (Zaks and Klibanov, 1984). Therefore, the suitability of a reaction medium is determined by the interaction of the solvent with water on the enzyme, and its ability to affect water–enzyme interaction. In the enzymatic transesterification of soybean oils with methanol and ethanol, Noureddini et al. (2005) reported that 40% less water was required to maintain enzyme activity when ethanol was used instead of methanol. This is mostly due to the lower hydrophobicity of methanol when compared to ethanol. Immobilized enzymes also differ in the way that they respond to an increase in hydration. This depends on the type of support used during immobilization.

On the other hand, from an environmental point of view, the use of organic solvents has to be minimized because of their negative environmental impact. In addition, the use of organic solvents is generally not acceptable in the food and pharmaceutical industries (Ikeda, 1992; Snyder et al., 1996). Furthermore, they are usually expensive and require separation from the reaction medium. The need for better technology with higher purity and environmentally friendly processes resulted in a search for alternatives. The best solvent to replace the organic solvent should dissolve substrates and at the same time avoid solvent separation. Therefore, green solvents, such as supercritical fluids and ionic liquids, have been put forward (Romero et al., 2005).

5.2   SUPERCRITICAL FLUIDS

5.2.1   SUPERCRITICAL TECHNOLOGY

The discovery of supercritical fluids (SCFs) was in 1822, when under supercritical conditions the phase boundary between the liquid and gaseous phases disappeared with an increase in temperature in a sealed glass vessel. However, it was not until 1869, when the existence of a supercritical state was discovered, that the term *critical point* was introduced. SCFs are defined as fluids at pressures and temperatures above their critical values. This, in turn, represents the highest values at which the vapor and liquid phases can coexist in equilibrium, as shown in the phase diagram of pure compound (Figure 5.1). When two or more phases coexist, the separation between the phases is distinct as a result of the differences in the properties of these phases. The phase diagram, shown in Figure 5.1, is commonly used to indicate the boundaries of solid, liquid, and gaseous regions. The melting–freezing curve is the equilibrium between solid and liquid phases, and the vaporization–condensation curve represents the coexistence of the liquid and gas phases. The sumlimation curve is the equilibrium between the solid and gaseous phases. These three curves intersect at a triple point, where all three phases coexist. In the supercritical state, the fluid can no longer be liquefied by raising the pressure, nor can gas be formed by raising the temperature. Table 5.2 lists values for critical temperatures and pressures of the different compounds that are commonly used in a supercritical state.

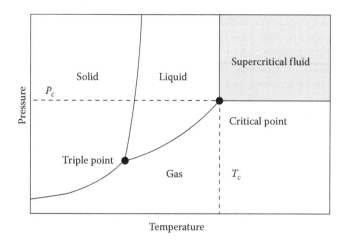

FIGURE 5.1   Pure component phase diagram. (From Liao, X., H. Zhang, and T. He, 2012, *Journal of Nanomaterials* 2012:12. With permission.)

TABLE 5.2
Critical Data of Some Common Substances Used in Supercritical States

Substance	Critical Temperature (°C)	Critical Pressure (bar)
Xenon	16.7	59.2
Carbon dioxide	31.1	72.8
Ethane	32.4	49.5
Nitrous oxide	36.6	73.4
Chlorodifluoromethane	96.3	50.3
Ammonia	132.4	115
Methanol	240.1	82
Water	374.4	224.1

Source: Data from Dean, J. R., 1998, *Extraction Methods for Environmental Analysis*, Vol. 3, John Wiley & Sons. With permission.

5.2.2 PHYSICOCHEMICAL PROPERTIES

Supercritical fluids (SCFs) and their associated technological uses have developed rapidly. They have been widely used in extraction and purification processes, as well as in organic synthesis processes (Manivannan and Sawan, 1998). The physical properties of SCFs are between those of gases and liquids. They have better transportation properties than liquids and better solvation properties than gases, and can be adjusted simply by changing the process pressure or temperature. A comparison between the physical properties of gases, liquids, and SCFs is illustrated in Table 5.3. The viscosity is similar to that of gases, which thus allows the fluid to flow easily through the openings. The diffusion is also lower than that of liquids, thus allowing fast transportation of the dissolved substances.

One of the key features of SCFs is their tunability to process operating conditions, which results in higher solvating powers. For example, the density of SCFs is sensitive to pressure. At a constant temperature, an increase in pressure results in an increase in fluid density. The relationship between pressure and density is shown in Figure 5.2, where

TABLE 5.3
Typical Ranges of Supercritical Properties

Property	Gas	SCF	Liquid
Density (gcm^{-3})	10^{-3}	$0.1-1$	1
Diffusivity (cm^2sec^{-1})	10^{-1}	$10^{-3}-10^{-4}$	10^{-5}
Dynamic viscosity ($gcm^{-1}sec^{-1}$)	10^{-4}	$10^{-3}-10^{-4}$	10^{-2}

Source: Data from Mukhopadhyay, M., 2000, *Natural Extracts Using Supercritical Carbon Dioxide*, Taylor & Francis. With permission.

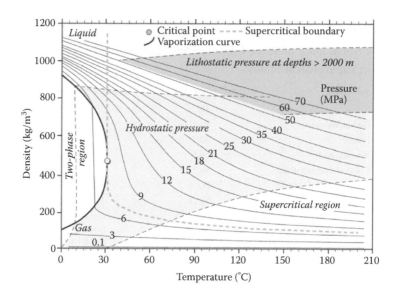

FIGURE 5.2   Density dependence of SC-CO₂ temperature and pressure. (From Bachu, S., 2003, *Environmental Geology* 44 (3):277–289. With permission.)

small changes in the pressure near the critical point can lead to a significant change in density, and around a fivefold increase in pressure results in significant changes in the density. Beyond 700 bar of pressure, however, the effect starts to diminish. This change in density affects other properties, such as the partition coefficient, dielectric constant, and the Hildebrand solubility parameter (Moriyoshi et al., 1993).

The operating conditions can also be changed to alter fluid hydrophobicity. This allows the fluid to solubilize a substance depending on the solute–fluid interaction without changing the fluid itself.

Many compounds have been used as SCFs. However, after the Montreal Protocol in 1987, the use of ozone-damaging compounds was banned and the use of SCFs was restricted to green compounds only. Among these solvents, carbon dioxide (CO_2), which has a critical temperature of 31.1°C and a critical pressure of 72.8 bar, is most commonly employed. CO_2 is nontoxic, nonflammable, noncorrosive, inexpensive, abundant, and an environmentally friendly compound.

5.2.3   SC-CO₂: Extraction Solvent

5.2.3.1   Process Features

Solvent extraction is the most common separation process. However, the use of organic solvents requires a solvent separation unit to separate it from the extract.

Considerable energy savings can be achieved by using a solvent that does not require separation. In that regard, supercritical carbon dioxide (SC-CO₂) is of great use in solute extraction. The ability of SC-CO₂ to extract a solute from a solid matrix depends mainly on molecular weight and polarity of the extract. For example, near the critical point, CO_2 acts as a good solvent for nonpolar to slightly polar solutes with low molecular weights (Berk, 2009). In comparison to organic solvent extraction, CO_2 is in a gaseous form at ambient conditions, which allows for complete separation and results in solvent-free extract, by depressurization. In addition, no oxidation or thermal degradation of the extract occurs.

Supercritical-CO_2 has been extensively studied in the extraction of essential oils (Reverchon and Marrone, 1997; Reverchon and Senatore, 1994), vegetable oils (Del Valle et al., 2004; Reverchon and Marrone, 2001; Sovova et al., 2001), and animal fats (Al-Zuhair et al., 2012; J. W. King et al., 1993; Taher et al., 2011) from a variety of sources. In addition, it has been used to extract astaxantine from *Haematococcus pluvialis* and phycocyanine from *Spirulina maxima* (Valderrama et al., 2003), β-carotene from *Chlorella vulgaris* (Kitada et al., 2009), and lutein from *Chlorella pyrenoidose* (Wu et al., 2007). It has been also used to extract lipids from *C. vulgaris* (Mendes et al., 1995, 2003), *C. cohnii* and *C. protothecoides* (Couto et al., 2010), *Nannochloropsis* sp. (Andrich et al., 2005), *S. platensis* (Andrich et al., 2006), *Chlorococum* sp. (Halim et al., 2011), and *S. maxima* (Mendes et al., 2003). It has been also used to extract fats from meat samples to produce lean fat meat with efficiency reaching 90% at 45°C, 500 bar and 3 ml min^{-1} (Taher et al., 2011). SC-CO_2 was found not to affect the texture and taste of the lean meat produced (Hashim et al., 2013). Therefore, it is a better alternative to organic solvents. Similar yields to *n*-hexane extraction were reported when SC-CO_2 was used to extract lipids from *S. platensis* (Andrich et al., 2006), *S. maxima* (Mendes et al., 2003), and *Pavlova* sp. (Cheng et al., 2011). The first plant using SC-CO_2 was established in 1970 in Germany for the decaffeination of coffee beans.

However, the use of nonpolar SC-CO_2 is not suitable for extracting polar solutes. To overcome this, the use of a co-solvent, also known as a modifier, in small amounts has been suggested to change the polarity of SC-CO_2 and increase its solvating power. For example, methanol has been used as a modifier for the extraction of nimbin from neem seeds (Tonthubthimthong et al., 2004), cocaine from human hair (Brewer et al., 2001), and pennyroyal essential oil (Aghel et al., 2004). Generally, the addition of the modifier increases the SC-CO_2 polarity and may increase matrix swelling. This enhances contact between the solute and the SC-CO_2.

5.2.3.2 Process Variables

Generally, extraction yields and extract characteristics are affected by the pretreatment of the raw material, solute solubility in SC-CO_2, and solvent flow rate. The selection of the best operating conditions for an efficient and cost-effective extraction is not an easy task and requires screening and reliable models for scaling-up from laboratory to pilot and industrial scales.

5.2.3.2.1 Pretreatment

Several sample characteristics have to be considered in solute extractions, such as moisture content, particle size, density, and porosity. The mass transfer area is an important parameter that should be taken in consideration. However, very small particles (powders) should always be avoided, as they may cause channeling inside the extraction bed and increase the drop in pressure. In addition, the sample should be relatively dry, since the use of a high water content sample can result in clogging.

5.2.3.2.2 Solubility

Solubility properties of the solute in SC-CO_2 depend on the extraction temperature and pressure, which are the main variables that affect the efficiency of the extraction process. It is well known that extraction yield increases with an increase in pressure if other factors are fixed, due to the increase in density. Temperature, however, has the opposite effects on extraction yields. An increase in temperature results in a reduction in fluid density that negatively affects the extraction yield. On the other hand, increasing the temperature also increases the solute vapor pressure, which enhances the solubility. At a crossover pressure, where the temperature does not show any balanced effect, the two competing effects are equal. At lower pressures than the crossover pressure, the change in density is predominant; at higher pressures the vapor pressure is predominant.

5.2.3.2.3 SC-CO_2 Flow Rate

Fluid flow rate is also considered in extraction optimization. It is usually used to determine whether the extraction is solubility or internal diffusion controlled. Typically, solubility controlled extractions show a direct correlation to the flow rate, whereas internal diffusion controlled extractions show this much less. In internal diffusion-controlled processes, the extraction yield can be increased by using smaller particles, as the specific area increases and the internal diffusion resistance lessens, due to a shorter diffusion path (Snyder et al., 1984). However, this is not always the case, as smaller particles may cause channeling (Eggers, 1996).

5.2.3.3 Extraction Kinetics and Modeling

To design and evaluate scale-up feasibility, the development of a reliable mathematical model to describe the process is important. In a packed bed extraction, the extraction fluid first penetrates and diffuses into the solid matrix. Then the solute solubilizes and moves from the solid matrix to the pores where diffusion inside the pores takes place. At the end, the fluid with the dissolved extract axially diffuses back along the extraction bed and leaves the extractor.

Most supercritical extractions are mass transfer processes with convection being the main transport mechanism. Therefore, accurate prediction of the convective mass transfer coefficient is important in describing process kinetics. Several approaches have been proposed to model overall extraction curves (OECs) of SCFs; each is based on certain postulations. They are mainly categorized into empirical models based on heat and mass analogs, and models based on differential mass balances. Empirical models that use exponential and hyperbolic correlations are the easiest. However, they are not adequate for scaling-up, as they do not give any information on mass transfer, which is important in designing a large-scale process (Huang et al., 2012). Bartle et al. (1990) developed the hot ball model, based on a heat transfer analogy, where the extraction process is treated as heat transfer phenomena, in which every particle was considered to be spherical and equations that covered the cooling of the hot balls were applied to describe the solute concentration profile.

The most commonly used mass transfer models, however, are those based on a differential mass balance of the solute over a control volume in the packed bed.

Several differential mass balance models have been proposed to characterize mass transfer kinetics. Among them, the model proposed by Sovova (1994), based on extraction from broken and intact cells (BICs), has been widely used. In this model, the solutes are stored in particle cells and protected by the cell wall. During the pretreatment step, to reduce the particle size and increase the surface area between the solute and fluid, some of the cells are broken and solutes become accessible to the fluid. These easily accessible solutes are denoted as x_p. The remaining solutes retained in unbroken cells are referred to as intact cells and defined as x_k. Thus, internal and external resistance control the extraction. Sovova (2005) also proposed a more complete model with additional parameters to consider equilibrium relationships. However, this modified model has not been widely used, due to its complexity, and most published work continues to use the older BIC model with mass transfer coefficients in the fluid (k_f) and solid (k_s) phases, and x_k as the main parameter. The following assumptions are usually considered:

- Isothermal and isobaric process conditions
- Constant physical properties of SC-CO$_2$ during the extraction
- Uniform fluid velocity during the extraction
- Uniform initial solute content and particle size distribution
- Constant bed porosity
- Negligible axial dispersion
- Negligible accumulation of solutes in SC-CO$_2$

The extraction of solutes from solid samples, according to the BIC model, can be divided into three periods: the constant extraction rate, falling extraction rate, and diffusion (Sovova, 1994). In the first period, the easily accessible solutes are extracted at a constant rate until the particles at the bed entrance lose all their accessible solutes. At this time, the diffusion extraction of the entrance of the bed begins, combined with convection extraction of the rest of the bed. This results in a continuation of the extraction for broken cells, beyond the bed entrance, and at the same time the extraction of cells begins at the bed entrance and the rate of extraction decreases. At the end of this stage, all of the broken cells are extracted and only the intact cells are left. Thus the extraction becomes diffusion controlled. According to this, the simplified mass balances of the fluid and solid phases are described by Equations 5.1 and 5.2, respectively:

$$u_{CO_2} \frac{\partial y}{\partial h} = \frac{j(x, y)}{\varepsilon} \frac{1}{\rho} \tag{5.1}$$

$$\frac{\partial x}{\partial t} = -\frac{j(x, y)}{(1-\varepsilon)} \frac{1}{\rho_s} \tag{5.2}$$

where u_{CO_2} is the interstitial velocity of SC-CO$_2$; h and H are the axial coordinate and bed height, respectively; x and y are the concentration of the solute in the solid and

SC-CO$_2$ phases, respectively; and ε, ρ, and ρ_s are the bed porosity, SC-CO$_2$ density, and solid density, respectively.

For the solvent-free solute at the entrance of the extraction bed, with solids that have the same initial solute concentration, the initial and boundary conditions are given in Equations 5.3 to 5.5:

$$x_{(h,t=0)} = x_o \tag{5.3}$$

$$y_{(h,t=0)} = y_{(h=0,t)} = 0 \tag{5.4}$$

$$x_{(h=o,t)} = 0 \tag{5.5}$$

The mass transfer flux, j, in the first extraction period, is controlled by the convection mass transfer and depends on solute concentration in the solid phase. It is expressed in Equation 5.6, whereas the flux inside the particles, which depends on the solute diffusion from the interior of the solid to the surface, is expressed by Equation 5.7:

$$j(x, y) = k_f a_o (Y^* - y), \quad \text{for} \quad x > x_k \tag{5.6}$$

$$j(x,y) = k_s a_o \left(1 - \frac{y}{Y^*}\right), \quad \text{for} \quad x \leq x_k \tag{5.7}$$

where Y^* is the solute solubility in SC-CO$_2$; k_f and k_s are mass transfer coefficients in fluid and solid phases, respectively; and a_o is the specific surface area.

An analytical solution, according to the Sovova model, for an OEC at each extraction period is given in terms of the amount of extract versus the specific amount of SC-CO$_2$ (q-dependent), and depends on the concentration of the initial solute (x_o) and the less accessible concentrations and parameters of Z and W. Equations 5.8 to 5.10 show the analytical solution at each stage. In the first stage, where $q < q_m$,

$$E = q \times Y^* \times [1 - \exp(-Z)] \tag{5.8}$$

In the second stage ($q_m \leq q < q_n$), the unbroken cells start to be extracted

$$E = Y^* \times [q - q_m \times \exp(Z_w - Z)] \tag{5.9}$$

In the last stage, $q \geq q_n$

$$E = x_o - \frac{Y^*}{W} \ln \left\{ \frac{1 + \left[\exp\left(W + \frac{x_o}{Y^*}\right) - 1\right]}{\exp\left[W \times (q_m - q) \times \frac{x_i}{x_o}\right]} \right\} \tag{5.10}$$

where

$$q = \frac{m_{CO_2}}{m_{bed}}, \quad \dot{q} = \frac{q}{t}, \quad \text{and} \quad q_m = \frac{x_o - x_i}{Y* \times Z} \tag{5.11}$$

$$q_n = q_m + \frac{1}{W} \ln \left(\frac{x_i + (x_o - x_i) \times \exp\left(\dfrac{W\,x_o}{Y*}\right)}{x_o} \right) \tag{5.12}$$

$$Z_w = \frac{Z \times Y*}{W \times x_o} \ln \left(\frac{x_o \times \exp[W \times (q - q_m)] - x_i}{x_o - x_i} \right) \tag{5.13}$$

The parameters Z and W are dimensionless parameters proportional to the fluid and solid phase mass transfer coefficients, respectively, according to Equations 5.14 and 5.15:

$$Z = \frac{k_f a_o \rho_f}{[\dot{q}(1-\varepsilon)\rho_s]} \tag{5.14}$$

$$W = \frac{k_s a_o}{[\dot{q}(1-\varepsilon)]} \tag{5.15}$$

where E is the amount of solute extracted; m_{CO_2} is the amount of SC-CO$_2$ passed; m_{bed} is the weight of the solute-free sample; and q and $\dot{q}$ are the specific mass and specific rate of SC-CO$_2$ per weight of solute-free cells passing through the extractor, respectively. q_m and q_n are the q-values representing the time when the extraction begins inside the particles and the easily accessible solutes are extracted, whereas Z_w is the dimensionless axial coordinate between the fast and slow extractions.

Presenting the process model as a mass transfer correlation is also common. This requires an understanding of the process's physical properties, namely, the density and viscosity of the SC-CO$_2$ and the mass diffusion of the solute in SC-CO$_2$. Dimensionless numbers, namely, Reynolds (Re) (Equation 5.16), which is related to fluid flow; Schmidt (Sc) (Equation 5.17), which is related to mass diffusivity; Grashof (Gr) (Equation 5.18), which is related to mass transfer via buoyancy forces due to difference in density difference between saturated SC-CO$_2$ with solute and pure SC-CO$_2$; and Sherwood (Sh) (Equation 5.19), which is related to mass transfer, are important in these correlations. In supercritical extraction, natural convection is not significant (Shi et al., 2007) and in this case, Sh$_F$ is related only to Re and Sc, as shown in Equation 5.19.

$$Re = \frac{\rho u_{CO_2} d_p}{\mu} \tag{5.16}$$

$$Sc = \frac{\mu}{\rho D} \tag{5.17}$$

$$Gr = d_p^3 \, g \, \Delta\rho \left(\frac{\rho}{\mu^2} \right) \tag{5.18}$$

$$Sh_F = \frac{k_f d_p}{D} = C_0 Re^{C_1} Sc^{C_2} \tag{5.19}$$

where g is the gravitational constant and $\Delta\rho$ is the difference in mixture density between the saturated SC-CO$_2$ with solute and pure SC-CO$_2$. This is found by using a Peng-Robinson equation with suitable mixing rules (Del Valle et al., 2006), and C_0, C_1, and C_2 as the adjustable parameters. Table 5.4 shows examples of proposed mass transfer correlations for SCF extraction when forced convection dominates. In every correlation, the exponent power for Sc is 0.33 and Re is between 0.5 and 0.8. However, negative values for these exponents have been reported by Mongkholkhajornsilp et al. (2005).

At a very low Reynolds number and when small particles are used, both forced and natural convection become important and need to be considered (Lim et al., 1989). In this case, the Sh_N function of Re, Sc, and Gr are shown in Equation 5.20:

$$Sh_N = C_3 (GrRe)^{C_4} \tag{5.20}$$

where Sh_N is the natural Sherwood number, and C_3 and C_4 are adjustable parameters. Churchill (1977) developed an equation combining forced and natural convection, as shown in Equation 5.21:

$$Sh = \sqrt[3]{Sh_F^3 + Sh_N^3} \tag{5.21}$$

TABLE 5.4

Common Correlations for Forced Convection Mass Transfer at Supercritical Conditions

Correlation	Applicability	References
$Sh_F = 0.38 \, Re^{0.83} \, Sc^{1/3}$	$2 \leq Sc \leq 20; 2 \leq Re \leq 20$	Tan et al., 1988
$Sh_F = 0.82 \, Re^{0.6} \, Sc^{1/3}$	$3 \leq Sc \leq 11; 1 \leq Re \leq 70$	M. B. King et al., 1993
$Sh_F = 0.2548 \, Re^{0.5} \, Sc^{1/3}$	$3 \leq Sc \leq 11; 1 \leq Re \leq 70$	King et al., 1997
$Sh_F = 0.206 \, Re^{0.8} \, Sc^{1/3}$	$Sc < 10; 10 \leq Re \leq 100$	Puiggene et al., 1997
$Sh_F = 3.173 \, Re^{-0.06} \, Sc^{-0.85}$	–	Mongkholkhajornsilp et al., 2005
$Sh_F = 0.085 \, Re^{-0.298} \, Sc^{1/3}$	–	Mongkholkhajornsilp et al., 2005
$Sh_F = 0.135 \, Re^{0.5} \, Sc^{1/3}$	–	Mongkholkhajornsilp et al., 2005

An attempt has been made to model the OECs of lipids from a microalgae bio-mass of *Scenedesmus* sp. strain using a BIC model (Taher et al., 2014). Figure 5.3 shows the experimental and BIC model of the OECs using SC-CO$_2$ at two different process scales. The model curves are shown as dotted lines. The results showed that the process was divided into three periods, as proposed by Sovova, with a good representation of experimental data (R^2 value equal to 97.5%). The dimensionless parameters (Z and W) and the intact cell concentrations (x_k) in the model were all determined by minimizing the errors between the experimental and predicted data. The initial values for Z values were estimated from the k_f values using the empirical correlation given by Tan et al. (1988). k_f and k_s were then re-evaluated, based on the Z and W determined values.

The solute solubility in SC-CO$_2$ can be determined either experimentally or empirically. In the experimental approach, the slope of the linear part of the extrac-tion curve, shown in Figure 5.3, at low flow rates represents the solubility of the solute. However, the second approach is based on models, such as the one suggested by Chrastil (1982) and shown in Equation 5.22:

$$Y^* = \rho^{k_o} \exp\left(a + \frac{b}{T}\right) \qquad (5.22)$$

where Y^* is solute solubility, ρ is SC-CO$_2$ density in kgm^{-3}, T is temperature in kel-vins, k_o is an association constant that describes the number of fluid molecules in the solvated complex formed between the solute and solvent molecules at equilibrium,

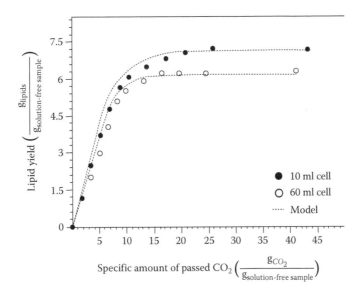

FIGURE 5.3 Experimental and predicted overall extraction curves for a 10 ml extraction cell and 60 ml extraction cell. (From Taher, H., S. Al-Zuhair, A. H. Al-Marzouqi, Y. Haik, and M. Farid, 2014, *Biomass and Bioenergy* 70:530–541. With permission.)

TABLE 5.5

Developed Mass Transfer Correlation of Lipid Extraction from *Scenedesmus* sp. Lipids Using SC-CO$_2$

Model	Strategy 1	Strategy 2
Forced	$Sh_F = 0.32\ Re^{1.2}\ Sc^{0.33}$	$Sh_N = 0.001\ (Gr\ Sc)^{0.56}$
Natural	$Sh_N = 6.3 \times 10^{-7}\ (Gr\ Sc)^{1.54}$	$Sh_F = 0.13\ Re^{1.4}\ Sc^{0.75}$
R^2	60	90

and a and b are constants related to the enthalpy of the solvation and solute molecular weights, respectively. This model, however, has some limitations in that it is not valid over a wide range of temperatures and for solubilities higher than 100 to 200 kgm^{-3} (Sparks et al., 2008). Thus, Adachi and Lu (1983) and Del Valle and Aguilera (1988) suggested some modifications, given in Equations 5.23 and 5.24:

$$Y* = \rho^{k_o + k_1\rho + k_2\rho^2} \exp\left(a + \frac{b}{T}\right) \tag{5.23}$$

$$Y* = \rho^{k_o} \exp\left(a + \frac{b}{T} + \frac{c}{T^2}\right) \tag{5.24}$$

where k_1, k_2, and c are parameters adjusted to the experimental data. The Adachi and Lu model considered the association parameters to be density dependent, whereas the Del Valle and Aguilera model considered that enthalpy changes with temperature, while keeping the association constant independent of the density.

The results were then used to develop a correlation that takes into consideration both forced and natural convection to account for low Re numbers. The k_f values were obtained from the OEC model and used to calculate the experimental Sh. Two fitting strategies were considered to determine models parameters. In the first strategy, which is commonly used, the exponent of the Sc number, C_2 in Equation 5.19, was fixed at 0.33, whereas in the second strategy, the exponent was left as an estimate. The two strategies showed different values for fitting parameters (Table 5.5), where strategy 2 showed a high R^2 value, exceeding 90%, as compared to only about 60% in strategy 1. In both strategies, the fitted parameters differed from those given in the literature and may have been the result of small particle sizes and, hence, Re. The validity of the mass correlations developed was also verified from their ability to predict large-scale behavior using the correlation parameters determined from the experimental data on a smaller scale.

5.2.3.4 Process Scale-Up

On an industrial scale, it is very important to produce exactly the same extract yield as in the small-scale system. Due to high capital investment in supercritical fluid extraction, the development of a reliable mathematical model that describes the process is necessary to design and evaluate scale-up feasibility. To guarantee the success of a large-scale extraction process, some scaling-up criteria have to be considered. Intensive properties that do not depend on quantities should always be fixed. This includes the extraction temperature, pressure, SC-CO_2 velocity, and bed characterization. Whereas, sample feed, solvent flow rate, and bed dimensions are affective parameters. As it is important to reproduce the same OECs on a large scale, Prado et al. (2011) suggested the use of simple criteria that could help in scaling up the process. The selection of the best strategy depends on limiting the extraction step. For example, in solubility controlled processes, the solvent-to-sample ratio $\left(\dfrac{S}{m} \right)$ should be kept constant between the two scales, and for processes that are diffusion controlled, the residence time, which is related to the solvent flow rate-to-sample ratio $\left(\dfrac{F}{m} \right)$ should be constant. When both steps are important, both ratios $\left(\dfrac{S}{m} \text{ and } \dfrac{F}{m} \right)$ should be kept constant.

This was successfully proven (Taher et al., 2014) in an attempt to scale up (eightfold) the SC-CO_2 extraction process of *Scenedesmus* sp. Lipids at 53°C and 200 bar were also kept constant in both small- and large-scale extractors. The correlations, shown in Figure 5.3, were used to predict the values of mass transfer coefficients on a larger scale. They were then used to predict the OECs. The extraction yield of the large-scale system was found to be slightly lower than that of the small-scale one. This was explained by the higher interstitial velocity (3-fold higher) and larger diameter (almost 2-fold) of the larger extraction cell when compared to the small cell. This had a negative effect on the yield.

5.2.4 SC-CO_2: Reaction Solvent

5.2.4.1 Process Features

In enzymatic reactions, SCFs have shown their potential as better alternatives to conventional organic solvents, due to easy product recovery, lower heating requirements, and side reactions. The first attempt to use SCFs in enzymatic reactions dates to 1985, when an alkaline phosphatase was used, followed by p-cresol and p-chlorophonol oxidation with polyphenol oxidase (Hammond et al., 1985). The fluids used in enzymatic reactions are those that have low critical temperatures to avoid the thermal denaturation of the enzyme. Therefore, CO_2, ethane, and fluoroform have been the commonly used. Supercritical water, however, cannot be used due to a high critical temperature (374.4°C), which deactivates the enzyme. On the other hand, it is well known that enzymes, and proteins in general, are not affected by high pressure (Lanza et al., 2004; Prasanth and Abraham, 2009).

Generally, SCFs play an important role in reactions controlled by mass transfer diffusion, rather than reaction kinetics. In this case, carrying out the reaction in a SCFs medium results in a faster reaction compared to one where the same reaction is carried out in an organic solvent medium at the same operating temperature. This is mainly due to the SCFs gaslike viscosities, which allows rapid solvent penetration into a solid matrix, high diffusion, and low surface tensions that improve substrates mass transfer into immobilized enzyme pores. Another important advantage of SCFs is the tunability of the fluid properties, hence the solvation power, simply by changing the pressure or the temperature. The adjustable properties of the fluid allow easy downstream separation of products and unreacted substrates. Cernia et al. (1998) and Celia et al. (1999) tested the performance of different lipases from different sources for biotransformations of different natural and synthetic substrates. Kumar et al. (2004) studied the esteratification of palmitic acid with ethanol using Novozym®435, Lipolase 100T, and hog pancreas lipase in solvent-free and SC-CO_2 systems. The reaction was superior in SC-CO_2 with high reaction rates at low enzyme loading and easy downstream processing.

Although, several solvents can be used as a medium in supercritical conditions, SC-CO_2 has been the most common. This is due to the low critical temperature of CO_2, as shown in Table 5.2, which is compatible with most enzymes. Supercritical ethane has also been used successfully in enzymatic-catalyzed reactions (Mesiano et al., 1999); however, it has never been used with lipase. The major disadvantage of SC-CO_2 is the nonpolarity of the fluid. However, this has recently been enhanced by surfactant developments that can improve the solubility of both hydrophilic and hydrophobic compounds in SC-CO_2 (Bender et al., 2008). The high cost of pumping to match the high pressure can be recovered by enzyme recycling and continuous production. As mentioned earlier, separation can be easily achieved by a pressure reduction, where the extracts do not dissolve in CO_2 at room temperature.

5.2.4.2   Process Variables

Several factors affect enzymatic reactions in SCFs. The solubility of the substrates and the activity and stability of the enzyme are the most common parameters that are usually considered. Pressure and temperature can always be used to change the reaction mixture density and other transport properties, which will significantly affect the solubility of the substrates and products in the fluid (Randolph et al., 1991). On the other hand, it may be difficult to understand the absolute effect of process variables and their ability to enhance enzyme stability and activity.

High pressures are required to reach a supercritical state. It was found that the stability and activity of many enzymes exposed to high-pressure fluids, like SC-CO_2, are either unaffected or, in some cases, enhanced. This is because high pressures affect enzyme confirmation and kinetics. For example, it was found that the initial reaction rate of isoamyl acetate synthesis using *C. antarctica* lipase B was higher in SC-CO_2 at 140 bar compared to in *n*-hexane at the same temperature and atmospheric pressure. However, the global esterification yield was similar for both solvents (Romero et al., 2005). The activity of *Pseudomonas fluorescens* and *Rhizopus oryzae* lipases was also significantly enhanced at high pressures (Chen et al., 2013). In the pressure range of 100 to 400 bar, which is typically used in lipase-catalyzed reactions, only reversible changes that do not negatively affect the activity of the enzyme can occur. Nevertheless, ultrahigh pressure may negatively affect enzyme activity and can lead to irreversible structural changes and denaturation.

The effect of pressure on reaction rate depends on its effect on the activation volumes, which is the difference in partial molar volume between the activated complex and reaction substrates. Celia et al. (2005) studied the effect of pressure on *P. cepacia* lipase activity in catalyzing the transesterification of 1-phenylethanol with a vinyl acetate in SC-CO_2. The reaction rate and enzyme selectivity were investigated by considering the activation volume of the reaction. It was found that by increasing the reaction pressure, the activation volume was reduced and thus resulted in an improved reaction rate. Typically, the activation volume can be either positive or negative. This indicates if the reaction is accelerated by pressure. The positive values show that the activated complex has a higher molar volume than do the substrates and that increasing the pressure would not favor the reaction, and vice versa. The effect of pressure on the reaction rate constant can be represented using Eyring's equation (Eyring et al., 1946), as shown in Equation 5.25. This is derived from the transition state theory.

$$\left(\frac{\partial \ln k}{\partial P} \right)_T = -\frac{\Delta V^*}{RT} \qquad (5.25)$$

where ΔV^* is the activation volume; k is kinetic constant; and R, T, and P are universal gas constant, reaction temperature, and pressure, respectively.

The enhancement of enzyme activity and stability in $SC\text{-}CO_2$ was demonstrated by several studies (de Oliveira Kuhn et al., 2011; Dhake et al., 2011; Manera et al., 2011). No change was observed with immobilized lipase from *Rhizomucor miehei* and native lipases from *P. fluorescens*, *Rhizopus javanicus*, *Rhizopus niveus*, and *Candida rugosa* when exposed to $SC\text{-}CO_2$ for one day at 40°C and 300 bar. Stability was also unaffected when stored for one month in $SC\text{-}CO_2$ at 40°C and 140 bar. A similar result was also reported with immobilized *Rhizomucor miehie* lipase used for an oleyl-oleate synthesis. This feature increases the feasibility of using immobilized lipases in high-pressure systems for industrial applications.

Another important factor that causes a decrease in the lipase activity is the depressurization rate, which is a step commonly used to separate the products. It has been reported that lipase activity decreases with an increase in the number of depressurizations (GieBauf et al., 1999). A rapid release of dissolved CO_2 in the water bound to the enzyme may cause structural change to the enzyme (Lin et al., 2006). This irreversible change with depressurization was studied by Randolph et al. (1991). However, this problem is only encountered in the batch system and results in enzyme inactivation. In continuous flow systems, however, the depressurization step takes place after the product leaves the reactor. Immobilized lipases are retained inside the reactor and not exposed to depressurization.

Although pressure may have an effect on enzyme activity, the effect of temperature is usually more significant. Enzymes are folded peptide molecules held together by hydrogen bonds, ionic interactions, and disulfide bridges. The latter are usually responsible for the stability of the enzymes (Kasche et al., 1988). When the enzymatic reaction mixture is exposed to a high temperature, the kinetic energy of the atoms that make up the enzyme are vigorously vibrated and this tears apart the hydrogen and other bonds holding the protein structure together. This results in enzyme denaturation. On the other hand, increasing the temperature increases the reaction rate, as it also increases the rate constant and lowers the reaction viscosity and mass transfer limitations. Therefore, there is an optimum temperature at which the reaction rate is maximized. The determination of an optimum temperature for an immobilized enzyme depends on the enzyme's nature and source. For example, Almeida et al. (1998) showed that at a constant SC-CO_2 pressure of 100 bar, an increase in reaction temperature from 35°C to 45°C resulted in an increase in the activity of Novozym®435 when used for the transesterification of n-butyl acetate by 1-hexanol.

In the presence of water, the activity of the lipase may be reduced in SC-CO_2 due to the formation of carbonic acid from the dissolution of the CO_2 in water. However, this may not be important, as many enzymes, including lipase from *Candida cylindracea*, maintain their activity at low pH values, resulting in some cases in a value of 2 (Yu et al., 1992). Another concern is that the surfaces of some enzymes, such as lipase, are decorated with lysine side chains. The SC-CO_2, as a Lewis acid, is likely to react with the amine groups, which are strong bases of these chains, resulting in a reversible formation of carbamates (Wright and Moore, 1948). The stability of the carbamate formed depends on SC-CO_2 properties and amine functionality. Generally, carbamates are stable only at low temperatures, but at high temperatures they become unstable, as the CO_2 is removed and the amine group is liberated.

5.3 IONIC LIQUIDS

5.3.1 CHARACTERISTICS AND PROPERTIES OF IONIC LIQUIDS

Recently, ionic liquids (ILs) have been put forward as green alternatives to conventional volatile organic solvents for different applications. ILs are organic salts that are composed of cations and anions, rather than molecules, and exist in a liquid state in ambient conditions with a low tendency toward crystallization. The first investigation of ILs was in 1914 when ethylammonium nitrate ([EtNH$_3$][NO$_3$]), which exists as a liquid at room temperature with a melting point of 12°C, was discovered when prepared by the neutralization of ethylamine with citric acid (Walden, 1914). However, this solvent did not receive much attention at that time due to its explosive nature. This was followed by the work of Hurley and Wier (1951), who were looking for alternative techniques to electroplate aluminum via heating. They prepared different mixtures of alkylpyridinium bromide with aluminum chloride ($AlCl_3$), and found that a mixture of N-ethylpyridinium bromide and $AlCl_3$, [C$_2$Py][AlBrCl$_3$], was liquid at room temperature. From these, ILs that are based on $AlCl_3$ have been put forward. However, due to the hygroscopic nature of $AlCl_3$ in liberating corrosive acids, the use of these ILs was also restricted.

Various cations and anions can be used to produce huge number of ILs. The most common are those composed of nitrogen or phosphorus-containing cations such as 1-alkyl-3-methylimidazolium, *N*-alkylpyridinium, *N,N*-dialkylpiperidinium, *N,N*-dialkylpyrrolidinium, tetraalkylammonium, or tetraalkylphosonium; and anions such as chloride, nitrate, hexafluorophosphate, tetrafluoroborate, ethyl sulfate, or bis-triflimide, as shown in Figure 5.4 and Table 5.6. The commercially available ILs are those based on 1-butyl-3-methylimidazolium with hexafluorophosphate and

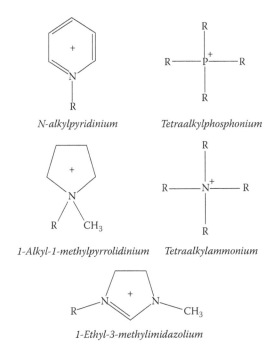

FIGURE 5.4   Most commonly used cations in IL combinations (R is the alkyl chain).

TABLE 5.6
Most Commonly Used Anions in IL Combinations

Anion	Full Name	Abbreviation
Cl⁻	Chloride	Cl
Br⁻	Bromide	Br
NO_3^-	Nitrate	NO_3
BF_4^-	Tetrafluoroborate	BF_4
PF_6^-	Hexafluorophosphate	PF_6
$(CF_3SO_2)_2N^-$	Bis(trifluoromethanesulfonyl)imide	Tf_2N
$CF_3SO_3^-$	Trifluorosulfonylimide	[TfO]
$C_4F_9SO_3^-$	Perfluorobutylsulfonate	[NfO]

tetrafluoroborate anions (Shariati and Peters, 2005; Shariati et al., 2005). The behavior of ILs are almost similar to those of organic solvents, therefore, they can replace volatile organic solvents in many applications, including in enzymatic processes (Earle and Seddon, 2000; Fan and Qian, 2010; Jain et al., 2005).

ILs are environmentally more compatible than organic solvents. Most important is their involatility, which arises from negligible vapor pressure. Therefore, they do not contaminate gas emissions and are characterized as green solvents. Other interesting properties are melting point, viscosity, and hydrophobicity. ILs also do not decompose over a large temperature range, which gives them a thermal-stability feature. In addition, they are able to dissolve many substrates, including lipids, and their properties can be tuned by varying the anion–cation combinations (Chinnappan et al., 2012; Holbrey and Seddon, 1999; Welton, 1999). Typically, ILs properties depend on the nature and structure of the cation and on the anion. By the judicious selection of the cation or anion component, these properties can be tailored. This is commonly referred to as solvent design.

5.3.1.1   Melting Point

The size and length of the cations or anions play an important role in determining the melting point (T_m) of ILs. Generally symmetric cations result in a higher T_m than those with asymmetric cations. For example, 1,3-dimethylimidazolium, which has a symmetric cation, has a higher melting point than ILs of asymmetric cations (Endres and El Abedin, 2006; Ohno and Yoshizawa, 2002). This is because symmetric cations are packed into a crystal structure, which is not the case with asymmetric ones. The melting point can also be affected by the size of the cations. For example, large cations such as in 1-butyl-3-methylimidazolium result in ILs with a lower T_m as compared to the smaller cations in 1-ethyl-3-methylimidazolium.

5.3.1.2   Viscosity

The viscosity of ILs is relatively high. It is determined by the nature of the interaction bonds, typically the Van der Waal's forces and hydrogen bonds. Values have been reported to be in the range of 10 to 1000 cP at room temperature (Wilkes, 2004). ILs with shorter alkyl chains on the cation show lower viscosity (Bonhôte et al., 1996; Huddleston et al., 2001; Kanatani et al., 2009; Mann et al., 2009) when compared to the ones with larger chains. In addition, a viscosity increase was also reported to be due to fluorination in cations/anions or electrostatic forces between charges and ions (Chinnappan et al., 2012; Tang et al., 2012).

5.3.1.3   Hydrophobicity

The miscibility of the solvent in water, referred to as hydrophobicity (log P), strongly depends on the nature of the cations and anions in the ILs. In that regard, ILs can be divided into water immiscible (hydrophobic) and water miscible (hydrophilic). The hydrophobicity is mainly determined by the choice of anion.

Hydrophilic anions such as Cl^-, Br^-, I^-, NO_3^-, CH_3COO^-, and CF_3COO^- generate hydrophilic ILs; and hydrophobic anions such as PF_6^- and Tf_2N^- generate hydrophobic ILs. The hydrophobicity can also be affected, but to a lesser extent, by the length of the alkyl chain in the cation. As the chain length increases, so does the hydrophobicity (Aki et al., 2001; Huddleston et al., 2001).

5.3.1.4   Density

The density of the ILs mainly depends on the length of the alkyl chain in the cation, and the interaction forces between the cation and the anion. ILs with short alkyl chains in the cation or anion are denser than those with lengthened alkyl chains (Huddleston and Rogers, 1998). Generally, ILs are denser than water.

5.3.2 LIPASE-CATALYZED REACTIONS IN IONIC LIQUIDS

The first attempt to use ILs as a reaction medium was in 2000 using *Candida antarctica* lipase in [bmim][PF$_6$] and [bmim][BF$_4$] (Madeira Lau et al., 2000). Since then the use of ILs in enzyme reactions has grown rapidly. Generally, two parameters are always considered when selecting a suitable IL: (1) the solubility of the reaction substrates and products in the IL, and (2) the interaction between the reaction medium and substrates. In addition, the effect of the ILs on the activity and stability of the lipase, either in free or immobilized form, should be considered.

The activity (Dang et al., 2007; de los Ríos et al., 2008; Eckstein et al., 2002; Lozano et al., 2002, 2003a,b; van Rantwijk and Sheldon, 2007), stability (Dang et al., 2007; Kaar et al., 2003; Klahn et al., 2011; Lozano et al., 2001, 2003a), and behavior (Madeira Lau et al., 2000) of many lipases in different ILs were found comparable to those in organic solvents. Madeira Lau et al. (2000) tested the use of *C. antarctica* lipase B in [bmim][PF$_6$] and [bmim][BF$_4$] in alcoholysis, ammoniolysis, and perhydrolysis reactions, and found comparable reaction rates to those in *tert*-butanol. Lozano et al. (2003a) tested ester synthesis using lipase in ILs based on dialkylimidazolium or quaternary ammonium cations associated with perfluorinated or bis(trifluoromethylsulfonyl)amide anions, and found that the activity of the lipase was enhanced when compared to that in organic solvents. de los Ríos, Hernández-Fernández, Martinez et al. (2007b) studied the transesterification of vinyl butyrate with 1-butanol catalyzed by *C. antarctica* lipase B for the synthesis of butyl butyrate in different imidazolium-based ILs. The activity and selectivity of the lipase in the water-immiscible ILs ([bmim][PF$_6$], [bdmim][PF$_6$], [hmim][PF$_6$], [omim][PF$_6$], [emim][NTf$_2$], [bmim][NTf$_2$], [hmim][NTf$_2$] and [omim][NTf$_2$]) were found to be higher than those obtained in *n*-hexane. *P. cepacia* lipase and *Candida rugosa* lipase also performed better than organic solvents. The ability of these two enzymes in [bmim][PF$_6$] were found to be better than in dichloromethane (Nara et al., 2002) and chloroform (Kim et al., 2003), respectively. Sheldon et al. (2002) studied the stability of *C. antarctica* lipase B in [bmim][PF$_6$] and found that both free and immobilized forms of the enzyme were stable. Similar results were also reported by Lozano et al. (2001).

5.3.3 CHALLENGES OF USING IONIC LIQUIDS

The negligible vapor pressure of ILs makes them good alternatives to more volatile solvents. However, such features can result in other problems, such as a difficulty in separating products from the solvent. Volatile products can be extracted from the ILs by distillation/evaporation. However, nonvolatile products cannot be easily separated (Zhao et al., 2005). The use of organic cosolvents can solve this problem. However, this contradicts the goal of using a green solvent. In addition, ILs are expensive to use at the industrial level, which is always associated with a purification step. Therefore, they need to be recycled for the process to be feasible. In addition, an IL that is composed of halogen anions such as [PF$_6$] or [BF$_4$] might be toxic due to their poor stability in water, resulting in the formation of corrosive acids (Cammarata et al., 2001; Imperato et al., 2007; Wasserscheid et al., 2002).

5.4   IL–SC-CO$_2$ BIPHASIC SYSTEMS

As mentioned earlier, although ILs show potential in enzyme-catalyzed reactions, product recovery remains the main challenge. Therefore, a process needs to be developed for further enhancement. The conventional product separation approaches are either distillation, which is energy intensive, or low temperature extraction using selective organic solvents, which requires another separation unit and adds to the overall process costs. Blanchard et al. (1999) suggested the use of SC-CO$_2$, which can carry the substrates to the IL and extract the products, instead of organic solvents. By utilizing depressurization the final product can be removed with no solvent residue.

Typically, the SC-CO$_2$ and IL system form a biphasic mixture that contains the enzyme in a denser phase, which is the IL phase; whereas the lighter phase acts as a carrier for the substrates and products. In enzyme-catalyzed reactions in a biphasic mixture, the immobilized enzymes are suspended in the IL phase, and reaction substrates are dissolved in SC-CO$_2$. The substrates diffuse from the bulk of the SC-CO$_2$ phase into the two-phase interface, followed by partitioning between the two phases and diffusion into the IL phase toward the active site of the enzyme. The products are then released in the IL phase and extracted by SC-CO$_2$ (de los Ríos et al. 2007a).

The extraction of the product from the ILs with SC-CO$_2$ is the most important advantage of the biphasic systems. Typically, the effectiveness of SC-CO$_2$ for extraction depends on the phase behavior of the binary IL–SC-CO$_2$ system. The solubility of CO$_2$ in the IL is important to allow for contact between the CO$_2$ and the products. The dissolved CO$_2$ also decreases the viscosity of the IL and therefore improves mass transfer. It has been reported that CO$_2$ is soluble in every IL, whereas ILs are not soluble in the gaseous CO$_2$ phase, even at high pressures (Blanchard et al., 2001). Such systems have been successfully used in the synthesis of esters.

Although the IL–SC-CO$_2$ system enhances the stability of the enzyme, a reduction in reaction rates was observed in the biphasic system due to portioning and mass transfer restrictions between the two phases. By adjusting the appropriate composition of cations and anions in the IL, the activity and selectivity of the enzyme, and the mass transfer between phases can be improved. However, as the reaction proceeds and more products are formed, the interaction between these products and the IL may cause a significant increase in the solubility of the IL in the SC-CO$_2$ (Wu et al., 2003, 2004).

The development of a continuous system, in which immobilized enzymes can be used and both solvents can be recycled, would decrease overall costs. Hernández et al. (2006) and de los Ríos et al. (2007b) developed a process combining an IL–SC-CO$_2$ system with membrane technology to synthesize vinyl propionate from vinyl propionate and 1-butanol at 50°C and 80 bar in a recirculating bioreactor with the presence of coated immobilized *C. antarctica* lipase B within the IL. It was found that the selectivity of the lipase increased with the use of ILs, compared to that achieved with supercritical carbon dioxide tested alone.

REFERENCES

Adachi, Y., and B. C. Y. Lu. 1983. "Supercritical Fluid Extraction with Carbon Dioxide and Ethylene." *Fluid Phase Equilibria* 14:147–156.

Adamczak, M., and S. H. Krishna. 2004. "Strategies for Improving Enzymes for Efficient Biocatalysis." *Food Technology and Biotechnology* 42 (4):251–264.

Aghel, N., Y. Yamini, A. Hadjiakhoondi, and S. M. Pourmortazavi. 2004. "Supercritical Carbon Dioxide Extraction of *Mentha Pulegium* L. Essential Oil." *Talanta* 62 (2):407–411.

Aki, S. N. V. K., J. F. Brennecke, and A. Samanta. 2001. "How Polar Are Room-Temperature Ionic Liquids?" *Chemical Communications* 5:413–414.

Almeida, M. C., R. Ruivo, C. Maia, L. Freire, T. C. de Sampaio, and S. Barreiros. 1998. "Novozym 435 Activity in Compressed Gases. Water Activity and Temperature Effects." *Enzyme and Microbial Technology* 22 (6):494–499.

Al-Zuhair, S., A. Hussein, A. H. Al-Marzouqi, and I. Hashim. 2012. "Continuous Production of Biodiesel from Fat Extracted from Lamb Meat in Supercritical CO_2 Media." *Biochemical Engineering Journal* 60:106–110.

Andrich, G., U. Nesti, F. Venturi, A. Zinnai, and R. Fiorentini. 2005. "Supercritical Fluid Extraction of Bioactive Lipids from the Microalga *Nannochloropsis* Sp." *European Journal of Lipid Science and Technology* 107 (6):381–386.

Andrich, G., A. Zinnai, U. Nesti, and F. Venturi. 2006. "Supercritical Fluid Extraction of Oil from Microalga *Spirulina (Arthrospira) Platensis.*" *Acta Alimentaria* 35 (2):195–203.

Bachu, S. 2003. "Screening and Ranking of Sedimentary Basins for Sequestration of CO_2 in Geological Media in Response to Climate Change." *Environmental Geology* 44 (3):277–289.

Bartle, K. D., A. A. Clifford, S. B. Hawthorne, J. J. Lagenfeld, D. J. Miller, and R. Robinson. 1990. "A Model for Dynamic Extraction Using a Supercritical Fluid." *Journal of Supercritical Fluids* 3 (3):143–149.

Bender, J. P., A. Junges, E. Franceschi, F. C. Corazza, C. Dariva, J. Vladimir Oliveira, and M. L. Corazza. 2008. "High-Pressure Cloud Point Data for the System Glycerol + Olive Oil + n-Butane + AOT." *Brazilian Journal of Chemical Engineering* 25:563–570.

Berk, Z. 2009. "Extraction." In *Food Process Engineering and Technology*, edited by Z. Berk, 259–277. Academic Press.

Blanchard, L. A., D. Hancu, E. J. Beckman, and J. F. Brennecke. 1999. "Green Processing Using Ionic Liquids and CO_2." *Nature* 399 (6731):28–29.

Blanchard, L. A., Z. Gu, and J. F. Brennecke. 2001. "High-Pressure Phase Behavior of Ionic Liquid/Co2 Systems." *Journal of Physical Chemistry B* 105 (12):2437–2444.

Bonhôte, P., A.-P. Dias, N. Papageorgiou, K. Kalyanasundaram, and M. Grätzel. 1996. "Hydrophobic, Highly Conductive Ambient-Temperature Molten Salts." *Inorganic Chemistry* 35 (5):1168–1178.

Brewer, W. E., R. C. Galipo, K. W. Sellers, and S. L. Morgan. 2001. "Analysis of Cocaine, Benzoylecgonine, Codeine, and Morphine in Hair by Supercritical Fluid Extraction with Carbon Dioxide Modified with Methanol." *Analytical Chemistry* 73 (11):2371–2376.

Cammarata, L., S. G. Kazarian, P. A. Salter, and T. Welton. 2001. "Molecular States of Water in Room Temperature Ionic Liquids." *Physical Chemistry Chemical Physics* 3 (23):5192–5200.

Celia, E. C., E. Cernia, I. D'Acquarica, C. Palocci, and S. Soro. 1999. "High Yield and Optical Purity in Biocatalysed Acylation of Trans-2-Phenyl-1-Cyclohexanol with *Candida Rugosa* Lipase in Non-Conventional Media." *Journal of Molecular Catalysis B: Enzymatic* 6 (5):495–503.

Celia, E. C., E. Cernia, C. Palocci, S. Soro, and T. Turchet. 2005. "Tuning *Pseudomonas Cepacea* Lipase (PCL) Activity in Supercritical Fluids." *Journal of Supercritical Fluids* 33 (2):193–199.

Cernia, E., C. Palocci, and S. Soro. 1998. "The Role of the Reaction Medium in Lipase-Catalyzed Esterifications and Transesterifications." *Chemistry and Physics of Lipids* 93 (1–2):157–168.

Chen, D., C. Peng, H. Zhang, and Y. Yan. 2013. "Assessment of Activities and Conformation of Lipases Treated with Sub- and Supercritical Carbon Dioxide." *Applied Biochemistry and Biotechnology* 169:2189–2201.

Cheng, C.-H., T.-B. Du, H.-C. Pi, S.-M. Jang, Y.-H. Lin, and H.-T. Lee. 2011. "Comparative Study of Lipid Extraction from Microalgae by Organic Solvent and Supercritical CO_2." *Bioresource Technology* 102 (21):10151–10153.

Chinnappan, A., H. Kim, C. Baskar, and I. T. Hwang. 2012. "Hydrogen Generation from the Hydrolysis of Sodium Borohydride with New Pyridinium Dicationic Salts Containing Transition Metal Complexes." *International Journal of Hydrogen Energy* 37 (13):10240–10248.

Chowdary, G. V., and S. G. Prapulla. 2002. "The Influence of Water Activity on the Lipase Catalyzed Synthesis of Butyl Butyrate by Transesterification." *Process Biochemistry* 38 (3):393–397.

Chrastil, J. 1982. "Solubility of Solids and Liquids in Supercritical Gases." *Journal of Physical Chemistry* 86 (15):3016–3021.

Churchill, S. W. 1977. "A Comprehensive Correlating Equation for Laminar, Assisting, Forced and Free Convection." *AIChE Journal* 23 (1):10–16.

Couto, R. M., P. C. Simões, A. Reis, T. L. Da Silva, V. H. Martins, and Y. Sánchez-Vicente. 2010. "Supercritical Fluid Extraction of Lipids from the Heterotrophic Microalga *Crypthecodinium Cohnii.*" *Engineering in Life Sciences* 10 (2):158–164.

Dang, D. T., S. H. Ha, S.-M. Lee, W.-J. Chang, and Y.-M. Koo. 2007. "Enhanced Activity and Stability of Ionic Liquid-Pretreated Lipase." *Journal of Molecular Catalysis B: Enzymatic* 45 (3–4):118–121.

de los Ríos, A. P., F. J. Hernández-Fernández, D. Gomez, M. Rubio, F. Tomás-Alonso, and G. Víllora. 2007a. "Understanding the Chemical Reaction and Mass-Transfer Phenomena in a Recirculating Enzymatic Membrane Reactor for Green Ester Synthesis in Ionic Liquid/Supercritical Carbon Dioxide Biphasic Systems." *Journal of Supercritical Fluids* 43 (2):303–309.

de los Ríos, A. P., F. J. Hernández-Fernández, F. A. Martinez, M. Rubio, and G. Víllora. 2007b. "The Effect of Ionic Liquid Media on Activity, Selectivity and Stability of *Candida Antarctica* Lipase B in Transesterification Reactions." *Biocatalysis and Biotransformation* 25 (2–4):151–156.

de los Ríos, A. P., F. J. Hernández-Fernández, F. Tomás-Alonso, D. Gómez, and G. Víllora. 2008. "Synthesis of Flavour Esters Using Free *Candida Antarctica* Lipase B in Ionic Liquids." *Flavour and Fragrance Journal* 23 (5):319–322.

de Oliveira Kuhn, G., C. Coghetto, H. Treichel, D. de Oliveira, and J. V. Oliveira. 2011. "Effect of Compressed Fluids Treatment on the Activity of Inulinase from *Kluyveromyces Marxianus* NRRL Y-7571 Immobilized in Montmorillonite." *Process Biochemistry* 46 (12):2286–2290.

Dean, J. R. 1998. *Extraction Methods for Environmental Analysis.* Vol. 3. John Wiley & Sons.

Degn, P., and W. Zimmermann. 2001. "Optimization of Carbohydrate Fatty Acid Ester Synthesis in Organic Media by a Lipase from *Candida Antarctica.*" *Biotechnology and Bioengineering* 74 (6):483–491.

Del Valle, J. M., and J. M. Aguilera. 1988. "An Improved Equation for Predicting the Solubility of Vegetable Oils in Supercritical Carbon Dioxide." *Industrial & Engineering Chemistry Research* 27 (8):1551–1553.

Del Valle, J. M., and J. C. De La Fuente. 2006. "Supercritical CO_2 Extraction of Oilseeds: Review of Kinetic and Equilibrium Models." *Critical Reviews in Food Science and Nutrition* 46 (2):131–160.

Del Valle, J. M., O. Rivera, M. Mattea, L. Ruetsch, J. Daghero, and A. Flores. 2004. "Supercritical CO_2 Processing of Pretreated Rosehip Seeds: Effect of Process Scale on Oil Extraction Kinetics." *Journal of Supercritical Fluids* 31 (2):159–174.

Dhake, K. P., K. M. Deshmukh, Y. P. Patil, R. S. Singhal, and B. M. Bhanage. 2011. "Improved Activity and Stability of Rhizopus Oryzae Lipase via Immobilization for Citronellol Ester Synthesis in Supercritical Carbon Dioxide." *Journal of Biotechnology* 156 (1):46–51.

Dordick, J. S. 1989. "Enzymatic Catalysis in Monophasic Organic Solvents." *Enzyme and Microbial Technology* 11 (4):194–211.

Earle, M. J., and K. R. Seddon. 2000. "Ionic Liquids: Green Solvents for the Future." *Pure and Applied Chemistry* 72 (7):1391–1398.

Eckstein, M., P. Wasserscheid, and U. Kragl. 2002. "Enhanced Enantioselectivity of Lipase from *Pseudomonas* Sp. at High Temperatures and Fixed Water Activity in the Ionic Liquid, 1-Butyl-3-Methylimidazolium Bis[(Trifluoromethyl)Sulfonyl]Amide." *Biotechnology Letters* 24 (10):763–767.

Eggers, R. 1996. "Supercritical Fluid Extraction (SFE) of Oilseeds/Lipids in Natural Products." In *Supercritical Fluid Technology in Oil and Lipid Chemistry*, edited by J. W. King and G. R. List, 35–64. American Oil Chemists' Society Press.

Eltaweel, M. A., R. N. Z. R. Abd Rahman, A. B. Salleh, and M. Basri. 2005. "An Organic Solvent-Stable Lipase from *Bacillus* Sp. Strain 42." *Annals of Microbiology* 55 (3):187–192.

Endres, F., and S. Z. El Abedin. 2006. "Air and Water Stable Ionic Liquids in Physical Chemistry." *Physical Chemistry Chemical Physics* 8 (18):2101–2116.

Eyring, H., F. H. Johnson, and R. L. Gensler. 1946. "Pressure and Reactivity of Proteins, with Particular Reference to Invertase." *Journal of Physical Chemistry* 50 (6): 453–464.

Fan, Y., and J. Qian. 2010. "Lipase Catalysis in Ionic Liquids/Supercritical Carbon Dioxide and Its Applications." *Journal of Molecular Catalysis B: Enzymatic* 66 (1–2):1–7.

GieBauf, A., W. Magor, D. J. Steinberger, and R. Marr. 1999. "A Study of Hydrolases Stability in Supercritical Carbon Dioxide (SC-CO_2)." *Enzyme and Microbial Technology* 24 (8–9):577–583.

Halim, R., B. Gladman, M. K. Danquah, and P. A. Webley. 2011. "Oil Extraction from Microalgae for Biodiesel Production." *Bioresource Technology* 102:178–185.

Halling, P. J. 1994. "Thermodynamic Predictions for Biocatalysis in Nonconventional Media: Theory, Tests, and Recommendations for Experimental Design and Analysis." *Enzyme and Microbial Technology* 16 (3):178–206.

Hammond, D. A., M. Karel, A. M. Klibanov, and V. J. Krukonis. 1985. "Enzymatic Reactions in Supercritical Gases." *Applied Biochemistry and Biotechnology* 11 (5):393–400.

Hashim, I., S. Al Nuaimi, H. Afifi, H. Taher, S. Al-Zuhair, and A. Al-Marzouqi. 2013. "Quality Characteristics of Low Fat Lamb Meat Produced by Supercritical Carbon Dioxide Extraction." *Global Journal of Biology, Agriculture and Health Sciences* 2:5–9.

Hernández, F. J., A. P. de los Ríos, D. Gómez, M. Rubio, and G. Víllora. 2006. "A New Recirculating Enzymatic Membrane Reactor for Ester Synthesis in Ionic Liquid/ Supercritical Carbon Dioxide Biphasic Systems." *Applied Catalysis B: Environmental* 67 (1–2):121–126.

Hernández-Rodríguez, B., J. Córdova, E. Bárzana, and E. Favela-Torres. 2009. "Effects of Organic Solvents on Activity and Stability of Lipases Produced by Thermotolerant Fungi in Solid-State Fermentation." *Journal of Molecular Catalysis B: Enzymatic* 61 (3–4):136–142.

Holbrey, J. D., and K. R. Seddon. 1999. "Ionic Liquids." *Clean Products and Processes* 1 (4):223–236.

Huang, Z., X.-H. Shi, and W.-J. Jiang. 2012. "Theoretical Models for Supercritical Fluid Extraction." *Journal of Chromatography A* 1250:2–26.

Huddleston, J. G., and R. D. Rogers. 1998. "Room Temperature Ionic Liquids as Novel Media for 'Clean' Liquid–Liquid Extraction." *Chemical Communications* (16):1765–1766.

Huddleston, J. G., A. E. Visser, W. M. Reichert, H. D. Willauer, G. A. Broker, and R. D. Rogers. 2001. "Characterization and Comparison of Hydrophilic and Hydrophobic Room Temperature Ionic Liquids Incorporating the Imidazolium Cation." *Green Chemistry* 3 (4):156–164.

Hurley, F. H., and T. P. Wier. 1951. "Electrodeposition of Metals from Fused Quaternary Ammonium Salts." *Journal of the Electrochemical Society* 98 (5):203–206.

Ikeda, M. 1992. "Public Health Problems of Organic Solvents." *Toxicology Letters* 64–65:191–201.

Imperato, G., B. König, and C. Chiappe. 2007. "Ionic Green Solvents from Renewable Resources." *European Journal of Organic Chemistry* 2007 (7):1049–1058.

Jain, N., A. Kumar, S. Chauhan, and S. M. S. Chauhan. 2005. "Chemical and Biochemical Transformations in Ionic Liquids." *Tetrahedron* 61 (5):1015–1060.

Kaar, J. L., A. M. Jesionowski, J. A. Berberich, R. Moulton, and A. J. Russell. 2003. "Impact of Ionic Liquid Physical Properties on Lipase Activity and Stability." *Journal of the American Chemical Society* 125 (14):4125–4131.

Kamat, S., J. Barrera, J. Beckman, and A. J. Russell. 1992. "Biocatalytic Synthesis of Acrylates in Organic Solvents and Supercritical Fluids: I. Optimization of Enzyme Environment." *Biotechnology and Bioengineering* 40 (1):158–166.

Kanatani, T., K. Matsumoto, and R. Hagiwara. 2009. "Syntheses and Physicochemical Properties of New Ionic Liquids Based on the Hexafluorouranate Anion." *Chemistry Letters* 38 (7):714–715.

Kasche, V., R. Schlothauer, and G. Brunner. 1988. "Enzyme Denaturation in Supercritical CO_2: Stabilizing Effect of S-S Bonds During the Depressurization Step." *Biotechnology Letters* 10 (8):569–574.

Khmelnitsky, Y. L., V. V. Mozhaev, A. B. Belova, M. V. Sergeeva, and K. Martinek. 1991. "Denaturation Capacity: A New Quantitative Criterion for Selection of Organic Solvents as Reaction Media in Biocatalysis." *European Journal of Biochemistry* 198 (1):31–41.

Kim, M.-J., M. Y. Choi, J. K. Lee, and Y. Ahn. 2003. "Enzymatic Selective Acylation of Glycosides in Ionic Liquids: Significantly Enhanced Reactivity and Regioselectivity." *Journal of Molecular Catalysis B: Enzymatic* 26 (3–6):115–118.

King, J. W., J. H. Johnson, W. L. Orton, F. K. McKeith, P. L. O'Connor, J. Novakofski, and T. R. Carr. 1993. "Fat and Cholesterol Content of Beef Patties as Affected by Supercritical CO_2 Extraction." *Journal of Food Science* 58 (5):950–952.

King, M. B., T. R. Bott, and O. Catchpole. 1993. "Physico-Chemical Data Required for the Design of Near-Critical Fluid Extraction Process." In *Extraction of Natural Products Using Near-Critical Solvents*, 184–231. Springer Netherlands.

King, J. W., M. Cygnarowicz-Provost, and F. Favati. 1997. "Supercritical Fluid Extraction of Evening Primrose Oil Kinetic and Mass Transfer Effects." *Italian Journal of Food Science* 9 (3):193–204.

Kitada, K., S. Machmudah, M. Sasaki, M. Goto, Y. Nakashima, S. Kumamoto, and T. Hasegawa. 2009. "Supercritical CO_2 Extraction of Pigment Components with Pharmaceutical Importance from *Chlorella Vulgaris*." *Journal of Chemical Technology & Biotechnology* 84 (5):657–661.

Klahn, M., G. S. Lim, and P. Wu. 2011. "How Ion Properties Determine the Stability of a Lipase Enzyme in Ionic Liquids: A Molecular Dynamics Study." *Physical Chemistry Chemical Physics* 13 (41):18647–18660.

Klibanov, A. M. 1995. "What Is Remembered and Why?" *Nature* 374 (6523):596–596.

Klibanov, A. M. 2001. "Improving Enzymes by Using Them in Organic Solvents." *Nature* 209:241–246.

Kumar, R., G. Madras, and J. Modak. 2004. "Enzymatic Synthesis of Ethyl Palmitate in Supercritical Carbon Dioxide." *Industrial & Engineering Chemistry Research* 43 (7):1568–1573.

Laane, C., S. Boeren, K. Vos, and C. Veeger. 1987. "Rules for Optimization of Biocatalysis in Organic Solvents." *Biotechnology and Bioengineering* 30 (1):81–87.

Lanza, M., W. L. Priamo, J. V. Oliveira, C. Dariva, and D. Oliveira. 2004. "The Effect of Temperature, Pressure, Exposure Time, and Depressurization Rate on Lipase Activity in $SCCO_2$." *Applied Biochemistry and Biotechnology* 113 (1–3):181–187.

Leitner, W., and P. G. Jessop, eds. 2014. *Green Solvents, Supercritical Solvents.* Vol. 4 of *Handbook of Green Chemistry,* edited by P. T. Anastas. John Wiley & Sons.

Li, L., W. Du, D. Liu, L. Wang, and Z. Li. 2006. "Lipase-Catalyzed Transesterification of Rapeseed Oils for Biodiesel Production with a Novel Organic Solvent as the Reaction Medium." *Journal of Molecular Catalysis B: Enzymatic* 43 (1–4):58–62

Liao, X., H. Zhang, and T. He. 2012. "Preparation of Porous Biodegradable Polymer and Its Nanocomposites by Supercritical CO_2 Foaming for Tissue Engineering." *Journal of Nanomaterials* 2012:12.

Lim, G. B., G. D. Holder, and Y. T. Shah. 1989. "Solid–Fluid Mass Transfer in a Packed Bed under Supercritical Conditions." In *Supercritical Fluid Science and Technology,* 379–395. American Chemical Society.

Lin, T.-J., S.-W. Chen, and A.-C. Chang. 2006. "Enrichment of N-3 PUFA Contents on Triglycerides of Fish Oil by Lipase-Catalyzed Trans-Esterification under Supercritical Conditions." *Biochemical Engineering Journal* 29 (1–2):27–34.

Lozano, P., T. de Diego, D. Carrié, M. Vaultier, and J. L. Iborra. 2001. "Over-Stabilization of *Candida Antarctica* Lipase B by Ionic Liquids in Ester Synthesis." *Biotechnology Letters* 23 (18):1529–1533.

Lozano, P., T. de Diego, D. Carrié, M. Vaultier, and J. L. Iborra. 2002. "Continuous Green Biocatalytic Processes Using Ionic Liquids and Supercritical Carbon Dioxide." *Chemical Communication* 7:692–693.

Lozano, P., T. de Diego, D. Carrié, M. Vaultier, and J. L. Iborra. 2003a. "Enzymatic Ester Synthesis in Ionic Liquids." *Journal of Molecular Catalysis B: Enzymatic* 21 (1–2):9–13.

Lozano, P., T. de Diego, D. Carrié, M. Vaultier, and J. L. Iborra. 2003b. "Lipase Catalysis in Ionic Liquids and Supercritical Carbon Dioxide at 150°C." *Biotechnology Progress* 19 (2):380–382.

Macrae, A. R. 1983. "Lipase-Catalyzed Interesterification of Oils and Fats." *Journal of the American Oil Chemists' Society* 60 (2):291–294.

Madeira Lau, R., F. Van Rantwijk, K. R. Seddon, and R. A. Sheldon. 2000. "Lipase-Catalyzed Reactions in Ionic Liquids." *Organic Letters* 2 (26):4189–4191.

Manera, A. P., G. Kuhn, A. Polloni, M. Marangoni, G. Zabot, S. J. Kalil, D. de Oliveira et al. 2011. "Effect of Compressed Fluids Treatment on the Activity, Stability and Enzymatic Reaction Performance of *B*-Galactosidase." *Food Chemistry* 125 (4):1235–1240.

Manivannan, G., and S. P. Sawan, 1998. "The Supercritical State." In *Supercritical Fluid Cleaning,* edited by John McHardy S. P. Sawan, 1–21. William Andrew.

Mann, J. P., A. McCluskey, and R. Atkin. 2009. "Activity and Thermal Stability of Lysozyme in Alkylammonium Formate Ionic Liquids-Influence of Cation Modification." *Green Chemistry* 11 (6):785–792.

Mendes, R. L., H. L. Fernandes, J. P. Coelho, E. C. Reis, J. M. S. Cabralb, J. M. Novais, and A. F. Palavra. 1995. "Supercritical CO_2 Extraction of Carotenoids and Other Lipids from *Chlorella Vulgaris*." *Food Chemistry* 53 (1):99–103.

Mendes, R. L., B. P. Nobre, M. T. Cardoso, A. P. Pereira, and A. F. Palavra. 2003. "Supercritical Carbon Dioxide Extraction of Compounds with Pharmaceutical Importance from Microalgae." *Inorganica Chimica Acta* 356:328–334.

Mesiano, A. J., E. J. Beckman, and A. J. Russell. 1999. "Supercritical Biocatalysis." *Chemical Reviews* 99 (2):623–634.

Miller, D. A., H. W. Blanch, and J. M. Prausnitz. 1991. "Enzyme-Catalyzed Interesterification of Triglycerides in Supercritical Carbon Dioxide." *Industrial & Engineering Chemistry Research* 30 (5):939–946.

Mongkholkhajornsilp, D., S. Douglas, P. L. Douglas, A. Elkamel, W. Teppaitoon, and S. Pongamphai. 2005. "Supercritical CO_2 Extraction of Nimbin from Neem Seeds— A Modelling Study." *Journal of Food Engineering* 71 (4):331–340.

Moriyoshi, T., T. Kita, and Y. Uosaki. 1993. "Static Relative Permittivity of Carbon Dioxide and Nitrous Oxide up to 30 Mpa." *Berichte der Bunsengesellschaft für physikalische Chemie* 97 (4):589–596.

Mukhopadhyay, M. 2000. *Natural Extracts Using Supercritical Carbon Dioxide*. Taylor & Francis.

Nakaya, H., O. Miyawaki, and K. Nakamura. 2001. "Determination of Log *P* for Pressurized Carbon Dioxide and Its Characterization as a Medium for Enzyme Reaction." *Enzyme and Microbial Technology* 28 (2–3):176–182.

Nara, S. J., J. R. Harjani, and M. M. Salunkhe. 2002. "Lipase-Catalysed Transesterification in Ionic Liquids and Organic Solvents: A Comparative Study." *Tetrahedron Letters* 43:2979–2982.

Nie, K., F. Xie, F. Wang, and T.I. Tan. 2006. "Lipase Catalyzed Methanolysis to Produce Biodiesel: Optimization of the Biodiesel Production." *Journal of Molecular Catalysis B: Enzymatic* 43 (1–4):142–147.

Noureddini, H., X. Gao, and R. S. Philkana. 2005. "Immobilized Pseudomonas Cepacia Lipase for Biodiesel Fuel Production from Soybean Oil." *Bioresource Technology* 96 (7):769–777.

Ohno, H., and M. Yoshizawa. 2002. "Ion Conductive Characteristics of Ionic Liquids Prepared by Neutralization of Alkylimidazoles." *Solid State Ionics* 154–155:303–309.

Parker, M. C., B. D. Moore, and A. J. Blacker. 1995. "Measuring Enzyme Hydration in Nonpolar Organic Solvents Using NMR." *Biotechnology and Bioengineering* 46 (5):452–458.

Prado, J. M., G. H. C. Prado, and M. A. A. Meireles. 2011. "Scale-Up Study of Supercritical Fluid Extraction Process for Clove and Sugarcane Residue." *Journal of Supercritical Fluids* 56 (3):231–237.

Prasanth, S., and T. E. Abraham. 2009. "Effect of Compression Pressure on the Activity of Lipase." *Hygeia Journal for Drugs and Medicines* 1 (1):11–18.

Puiggene, J., M. A. Larrayoz, and F. Recasens. 1997. "Free Liquid-to-Supercritical Fluid Mass Transfer in Packed Beds." *Chemical Engineering Science* 52 (2):195–212.

Radzi, S., M. Basri, A. Salleh, A. Ariff, R. Mohammad, M. B. Abdul Rahman, and R. N. Z. R. Abdul Rahman. 2005. "High Performance Enzymatic Synthesis of Oleyl Oleate Using Immobilised Lipase from *Candida Antartica*." *Electronic Journal of Biotechnology* 8 (3).

Randolph, T. W., H. W. Blanch, and D. S. Clark. 1991. "Biocatalysis in Supercritical Fluids." In *Biocatalysts for Industry*, edited by Jonathan S. Dordick, 219–237. Springer.

Reverchon, E., and F. Senatore. 1994. "Supercritical Carbon Dioxide Extraction of Chamomile Essential Oil and Its Analysis by Gas Chromatography-Mass Spectrometry." *Journal of Agricultural and Food Chemistry* 42 (1):154–158.

Reverchon, E., and C. Marrone. 1997. "Supercritical Extraction of Clove Bud Essential Oil: Isolation and Mathematical Modeling." *Chemical Engineering Science* 52 (20): 3421–3428.

Reverchon, E., and C. Marrone. 2001. "Modeling and Simulation of the Supercritical CO_2 Extraction of Vegetable Oils." *Journal of Supercritical Fluids* 19 (2):161–175.

Romero, M. D., L. Calvo, C. Alba, M. Habulin, M. Primozic, and Z. Knez. 2005. "Enzymatic Synthesis of Isoamyl Acetate with Immobilized Candida Antarctica Lipase in Supercritical Carbon Dioxide." *Journal of Supercritical Fluids* 33 (1):77–84.

Sabeder, S., M. Habulin, and Z. Knez. 2006. "Lipase-Catalyzed Synthesis of Fatty Acid Fructose Esters." *Journal of Food Engineering* 77 (4):880–886.

Shariati, A., and C. J. Peters. 2005. "High-Pressure Phase Equilibria of Systems with Ionic Liquids." *Journal of Supercritical Fluids* 34 (2):171–176.

Shariati, A., K. Gutkowski, and C. J. Peters. 2005. "Comparison of the Phase Behavior of Some Selected Binary Systems with Ionic Liquids." *AIChE Journal* 51 (5):1532–1540.

Sheldon, R. A., R. Madeira Lau, M. J. Sorgedrager, F. van Rantwijk, and K. R. Seddon. 2002. "Biocatalysis in Ionic Liquids." *Green Chemistry* 4 (2):147–151.

Shi, J., Y. Kakuda, X. Zhou, G. Mittal, and Q. Pan. 2007. "Correlation of Mass Transfer Coefficient in the Extraction of Plant Oil in a Fixed Bed for Supercritical CO_2." *Journal of Food Engineering* 78 (1):33–40.

Shimada, Y., Y. Watanabe, T. Samukawa, A. Sugihara, H. Noda, H. Fukuda, and Y. Tominaga. 1999. "Conversion of Vegetable Oil to Biodiesel Using Immobilized *Candida Antarctica* Lipase." *Journal of the American Oil Chemists' Society* 76 (7):789–793.

Snyder, J. M., J. P. Friedrich, and D. D. Christianson. 1984. "Effect of Moisture and Particle Size on the Extractability of Oils from Seeds with Supercritical CO_2." *Journal of American Oil Chemists' Society* 61 (12):1851–1856.

Snyder, J. M., J. W. King, and M. A. Jackson. 1996. "Fat Content for Nutritional Labeling by Supercritical Fluid Extraction and an On-Line Lipase Catalyzed Reaction." *Journal of Chromatography A* 750 (1–2):201–207.

Soumanou, M. M., and U. T. Bornscheuer. 2003. "Improvement in Lipase-Catalyzed Synthesis of Fatty Acid Methyl Esters from Sunflower Oil." *Enzyme and Microbial Technology* 33 (1):97–103.

Sovova, H. 1994. "Rate of the Vegetable Oil Extraction with Supercritical CO_2—I. Modelling of Extraction Curves." *Chemical Engineering Science* 49 (3):409–414.

Sovova, H. 2005. "Mathematical Model for Supercritical Fluid Extraction of Natural Products and Extraction Curve Evaluation." *The Journal of Supercritical Fluids* 33 (1):35–52.

Sovova, H., M. Zarevucka, M. Vacek, and K. Stransky. 2001. "Solubility of Two Vegetable Oils in Supercritical CO_2." *Journal of Supercritical Fluids* 20 (1):15–28.

Sparks, D. L., R. Hernandez, and L. A. Estevez. 2008. "Evaluation of Density-Based Models for the Solubility of Solids in Supercritical Carbon Dioxide and Formulation of a New Model." *Chemical Engineering Science* 63 (17):4292–4301.

Taher, H., S. Al-Zuhair, A. Al-Marzouqui, and I. Hashim. 2011. "Extracted Fat from Lamb Meat by Supercritical CO_2 as Feedstock for Biodiesel Production." *Biochemical Engineering Journal* 55 (1):23–31.

Taher, H., S. Al-Zuhair, A. H. Al-Marzouqi, Y. Haik, and M. Farid. 2014. "Mass Transfer Modeling of *Scenedesmus* Sp. Lipids Extracted by Supercritical CO_2." *Biomass and Bioenergy* 70:530–541.

Tan, C.-S., S.-K. Liang, and D.-C. Liou. 1988. "Fluid-Solid Mass Transfer in a Supercritical Fluid Extractor." *The Chemical Engineering Journal* 38 (1):17–22.

Tang, S., G. A. Baker, and H. Zhao. 2012. "Ether- and Alcohol-Functionalized Task-Specific Ionic Liquids: Attractive Properties and Applications." *Chemical Society Reviews* 41 (10):4030–4066.

Tonthubthimthong, P., P. L. Douglas, S. Douglas, W. Luewisutthichat, W. Teppaitoona, and L. Pengsopa. 2004. "Extraction of Nimbin from Neem Seeds Using Supercritical CO_2 and a Supercritical CO_2-Methanol Mixture." *Journal of Supercritical Fluids* 30 (3):287–301.

Tsitsimpikou, C., H. Daflos, and F. N. Kolisis. 1997. "Comparative Studies on the Sugar Esters Synthesis Catalysed by *Candida Antarctica* and *Candida Rugosa* Lipases in Hexane." *Journal of Molecular Catalysis B: Enzymatic* 3 (1–4):189–192.

Valderrama, J. O., M. Perrut, and W. Majewski. 2003. "Extraction of Astaxantine and Phycocyanine from Microalgae with Supercritical Carbon Dioxide." *Journal of Chemical & Engineering Data* 48 (4):827–830.

van Rantwijk, F., and R. A. Sheldon. 2007. "Biocatalysis in Ionic Liquids." *Chemical Reviews* 107 (6):2757–2785.

Walden, P. 1914. "Molecular Weights and Electrical Conductivity of Several Fused Salts." *Bulletin of the Russian Academy of Sciences*, 405–422.

Wasserscheid, P., R. van Hal, and A. Bosmann. 2002. "1-N-Butyl-3-Methylimidazolium ([Bmim]) Octylsulfate-an Even 'Greener' Ionic Liquid." *Green Chemistry* 4 (4):400–404.

Welton, T. 1999. "Room-Temperature Ionic Liquids. Solvents for Synthesis and Catalysis." *Chemical Reviews* 99 (8):2071–2084.

Wilkes, J. S. 2004. "Properties of Ionic Liquid Solvents for Catalysis." *Journal of Molecular Catalysis A: Chemical* 214 (1):11–17.

Wright, H. B., and M. B. Moore. 1948. "Reactions of Aralkyl Amines with Carbon Dioxide." *Journal of the American Chemical Society* 70 (11):3865–3866.

Wu, W., J. Zhang, B. Han, J. Chen, Z. Liu, T. Jiang, J. Hea, and W. Li. 2003. "Solubility of Room-Temperature Ionic Liquid in Supercritical CO_2 with and without Organic Compounds." *Chemical Communications* (12):1412–1413.

Wu, W., W. Li, B. Han, T. Jiang, D. Shen, Z. Zhang, D. Sun, and B. Wang. 2004. "Effect of Organic Cosolvents on the Solubility of Ionic Liquids in Supercritical CO_2." *Journal of Chemical & Engineering Data* 49 (6):1597–1601.

Wu, Z., S. Wu, and X. Shi. 2007. "Supercritical Fluid Extraction and Determination of Lutein in Heterotrophically Cultivated *Chlorella Pyrenoidosa*." *Journal of Food Process Engineering* 30 (2):174–185.

Yu, Z. R., S. S. H. Rizvi, and J. A. Zollweg. 1992. "Enzymic Esterification of Fatty Acid Mixtures from Milk Fat and Anhydrous Milk Fat with Canola Oil in Supercritical Carbon Dioxide." *Biotechnology Progress* 8 (6):508–513.

Zaks, A., and A. M. Klibanov. 1984. "Enzymatic Catalysis in Organic Media at 100 Degrees C." *Science* 224:1249–1251.

Zaks, A., and A. M. Klibanov. 1988. "The Effect of Water on Enzyme Action in Organic Media." *Journal of Biological Chemistry* 263 (17):8017–8021.

Zhao, H., S. Xia, and P. Ma. 2005. "Use of Ionic Liquids as 'Green' Solvents for Extractions." *Journal of Chemical Technology & Biotechnology* 80 (10):1089–1096.

6 Lipase-Catalyzed Production of Biodiesel Using Supercritical Technology

Currently, the world's energy demands are met mainly by fossil fuels. These resources are dwindling, resulting in an increase in their price. At the same time, the global environment is being threatened by an increase in harmful gas emissions associated with the overconsumption and combustion of fossil fuels. The transportation sector is one of the main fossil fuel consumers and also the biggest contributor to global environmental pollution. These challenges call for the development of alternative renewable and sustainable fuels that can meet increased energy needs. Biodiesel is potentially an acceptable and suitable substitute. This chapter presents technological breakthroughs in biodiesel production, its possible feedstocks, and production technologies. Transesterification of lipids from different feedstocks, in both chemically and lipase-catalyzed processes, is discussed. The implementation and also limitations of supercritical carbon dioxide (SC-CO$_2$) is considered in lipase-catalyzed transesterification processes. The proposed employment of supercritical fluids (SCFs) in an integrated continuous extraction–reaction process is also presented.

6.1 BIODIESEL

Biodiesel, a mixture of alkyl esters, has received increasing attention on a global scale (Bajpai and Tyagi, 2006; Demirbas, 2007a). It has many advantages over petroleum fuels. For example, it is renewable, nontoxic, biodegradable, and does not contribute to the net accumulation of greenhouse gases (Bajpai and Tyagi, 2006; Fellows, 2000). The physical and chemical properties of biodiesel are quite similar to those of petroleum diesel. Thus, biodiesel has been suggested as a possible replacement for transportation fuels and can be used without the need for any modification of existing engine technology (Demirbas, 2009b; Fjerbaek et al., 2009). Also, biodiesel has no sulfur content, a lower aromatic content (80% to 90% less), and a higher cetane number and flash point than petroleum diesel (Al-Zuhair, 2007; Demirbas, 2007a; Fukuda et al., 2001; Ranganathan et al., 2008). Other benefits of biodiesel include improved lubricity and lower emissions of certain harmful gases (Demirbas, 2007a). When compared to petroleum diesel, a 42.7% to 47.5% reduction in carbon monoxide and a 55.3% to 66.7% reduction in particulate matter emissions were reported (Demirbas, 2007a; Schumacher et al., 2001). However, when using biodiesel, higher levels of nitrogen oxide (by up to 11.5%) have been reported (Hess et al., 2005). Nevertheless, this can be improved by modifications in combustion temperature and injection timing (Hess et al., 2005; Wang et al., 2000). Therefore, biodiesel will not only decrease dependency on fossil fuels but will also significantly reduce greenhouse gases and other pollutants.

6.2 FEEDSTOCKS

Biodiesel can be produced from a variety of feedstocks with high heat content. This includes most vegetable oils and animal fats. They can also be produced from other sources like nonedible oils, waste cooking oils, grease, and algae oil (Demirbas, 2007b). The fuel's properties are important and depend on the properties and composition of the feedstock lipids; mainly on the fatty acids present in the oil. Typical fatty acids in these feedstocks are shown in Table 6.1, where lauric acid (C12:0), myristic acid (C14:0), palmitic acid (C16:0), stearic acid (C18:0), oleic acid (C18:1), and linolenic acid (C18:3) are predominant. High concentrations of saturated fatty acids and low levels of polyunsaturated fatty acids usually result in better ignition properties, engine performance, and oxidation stability.

Generally, the fuel properties of biodiesel can be characterized by determining the heating value, cetane number, cloud and pour points, flash point, sulfur content, acid value, and flow properties. Several standards have been revised for fuel properties. Among these, ASTM D6751 and EN14214, as given in Table 6.2, are the most common. A typical measure for biodiesel's energy content is derived from the heating value, also known as the caloric value, which is obtained from a complete combustion of the fuel source in an adiabatic bomb calorimeter under certain well-defined conditions. This usually ranges from 37.27 to 40.48 MJ kg^{-1} (Demirbas, 1998). It can also be calculated from lipid properties, such as iodine and saponification values (Demirbas, 1998); from fatty acid profiles (Fassinou et al., 2010; Knothe, 2005, 2009; Ramos et al., 2009); or from proximate analysis, through fixed carbon, volatile matter, and ash contents determination (Yin, 2011). The delay in ignition time between fuel injection and ignition is the cetane number. The higher the cetane number, the shorter the ignition delay time. Most biodiesels produced from vegetable oils have cetane numbers higher than 50 (Azam et al., 2005; Graboski and McCormick, 1998), however, the exact value depends on the fatty acid composition of the oil.

In cold environments, the cloud and pour points of the fuel are significant. The cloud point is the temperature at which wax formation begins to block the fuel filter, and the pour point is the temperature at which fuel is no longer pumped. The flash point is the temperature to which the fuel must be heated to achieve a flammable mixture of vapor and air that can be ignited. The higher the flash point, the lower the risk of firing. This is, therefore, safer with regard to the storage and transportation of the fuel.

Another important fuel property is viscosity, which measures the fuel's flow resistance. Higher viscosity lowers the performance, as is the case with the direct use of vegetable oils. In addition, biodiesel is subjected to oxidation, which might lead

TABLE 6.1

Fatty Acid Composition of Lipids from Common Feedstocks Tested for Biodiesel Production

Fatty Acids	Fatty Acid Composition (wt%)										
	$C_{14:0}$	$C_{16:0}$	$C_{16:1}$	$C_{18:0}$	$C_{18:1}$	$C_{18:2}$	$C_{18:3}$	$C_{20:0}$	$C_{20:1}$	$C_{22:0}$	Others
Vegetable Oils											
Canola oil	–	4	–	2	61	22	10	–	1	–	–
Corn oil	–	11	–	2	28	58	1	–	–	–	–
Cotton oil	1	23	1	2	17	56	–	–	–	–	–
Groundnut oil	–	11.2	–	3.6	41.1	35.5	0.1	–	–	–	–
Linseed oil	–	5.6	–	3.2	17.7	15.7	57.8	–	–	–	–
Olive oil	–	13.8	1.4	2.8	71.6	9	1	–	–	–	–
Palm oil	1	45	–	4	39	11	–	–	–	–	–
Sesame oil	–	9.6	0.2	6.7	41.1	41.2	0.7	–	–	–	–
Soybean oil	–	11	–	4	23	54	8	–	–	–	–
Sunflower oil	–	6	–	5	29	58	1	–	–	1	–
Animal Fats											
Beef fat	4	26	4	20	28	3		–	–	–	14
Chicken fat	1	25	8	6	41	18	1	–	–	–	–
Lamb fat	2	19		26	44	2	4	–	–	–	–

(Continued)

TABLE 6.1 (CONTINUED)
Fatty Acid Composition of Lipids from Common Feedstocks Tested for Biodiesel Production

Fatty Acids	Fatty Acid Composition (wt%)										
	$C_{14:0}$	$C_{16:0}$	$C_{16:1}$	$C_{18:0}$	$C_{18:1}$	$C_{18:2}$	$C_{18:3}$	$C_{20:0}$	$C_{20:1}$	$C_{22:0}$	Others
Madhuca longifolia	–	18	–	14	46	18	–	–	–	–	–
Pongamia pinnata	–	9	–	8	66	12	–	1	1	3	–
Microalgae Lipids											
Spirulina platensis	0.7	45.5	9.6	1.3	3.8	14.5	21.4	–	–	–	3.2
Spirulina maxima	0.3	45.1	6.8	1.4	1.9	14.6	20.6	–	–	–	9.3
Scenedesmus obliquus	0.6	16	8	0.3	8	6	28	–	–	–	33.1
Chlorella vulgaris	0.9	20.4	5.8	15.3	6.6	1.5	–	–	–	–	49.5
Dunaliella bardawil	–	41.7	7.3	2.9	8.8	15.1	20.5	–	–	–	3.7

Source: Becker, W., 2004, "Microalgae in Human and Animal Nutrition," in *Handbook of Microalgal Culture: Biotechnology and Applied Phycology*, edited by A. Richmond, Blackwell Science; Demirel, G. et al., 2006, *Meat Science* 72 (2):229–235; Kamal-Eldin, A., and R. Andersson, 1997, *Journal of the American Oil Chemists' Society* 74 (4):375–380; Sharma, Y. C., and B. Singh, 2010, *Fuel Processing Technology* 91 (10):1267–1273; and Sharma, Y. C. et al., 2010, *Energy and Fuels* 24 (5):3223–3231.

TABLE 6.2
Fuel Standards and Test Methods for Pure Biodiesel

Property	Method		Limit	
	ASTM	EN	ASTM	EN
Acid value (mgKOH g⁻¹)	ASTM D664	EN 14104	0.5 max	0.5 max
Water and sediment	ASTM D2709	EN ISO 12937	0.05 vol% max	500 mg kg⁻¹ max
Ester content	–	EN 14103	–	96.5 mol% min
Free glycerol	ASTM	EN 14105	0.02 wt%	0.02 mol%
Total glycerol	ASTM	EN 14106	0.24 wt%	0.25 mol%
Sulfur content	ASTM D874	ISO 3897	0.02 wt%	0.02 mol%
Phosphorous content	ASTM D4951	EN 14107	0.001 wt% max	10 mg kg⁻¹ max
Cetane number	ASTM D613	EN ISO 5165	47.0 min	51.0 min
Oxidation stability (h)	–	EN 14112	3 min	6 min
Flash point (°C)	ASTM D93	EN ISO 3679	93 min	120
Density (kg m⁻³, 15°C)	–	EN ISO 3675	–	860–900
Kinematic viscosity (mm² s⁻¹, 40°C)	ASM D445	EN ISO 3104	1.9–6.0	3.5–5.0

to degradation. Thus, it is crucial to determine fuel resistance to chemical changes brought about by oxidation.

6.2.1 VEGETABLE OILS

Since vegetable seeds are available in large quantities, they have been conventionally used for biodiesel production. These include soybean (Antunes et al., 2008; Liu et al., 2008), canola (Dubé et al., 2007), palm (Abdullah et al., 2009; Al-Zuhair et al., 2007; Kalam and Masjuki, 2002), and sunflower (Georgogianni et al., 2008; Pereyra-Irujo et al., 2009) seeds, which contain natural oils that can extracted and converted to biodiesel. However, the use of these edible oils for fuel production competes with their uses as food and increases food prices. In addition, the use of crops requires the development of agricultural land, and the need for fertilization and fresh water for irrigation. Moreover, the price of vegetable oils is high, and it has been reported that 60% to 75% of biodiesel costs comes from the cost of the vegetable oils used in their production (Al-Zuhair, 2007; Lai et al., 2005).

6.2.2 NONEDIBLE OILS

To overcome the problem of using feedstock that can be used as food, nonedible oils such as those from jatropha (*Jatropha curcas*), karanja (*Milletia pinnata*), caster (*Ricinus communis*), mahua (*Madhuca longifolia*), rubber (*Ficus elastica*), polanga (*Calophyllum inophyllum*), and tobacco (*Nicotiana tabacum*) have been suggested (Balat, 2011; Berchmans and Hirata, 2008; Ghadge and Raheman, 2005; Karmee and Chadha, 2005; Koh and Ghazi, 2011). Due to the presence of toxic compounds, such as curcin in *Jatropha curcas*, these feedstocks are not suitable for human consumption. It was believed that the use of such oils for fuel production might also help to solve food security problems. However, this alternative does not completely solve the problem, as plantations of nonedible oils require agricultural land, fertilization, manpower, and fresh water.

6.2.3 ANIMAL FATS AND WASTE COOKING OILS

Many animal processing and rendering companies create large amounts of waste fats. These differ from vegetable oils in their fatty acid composition; vegetable oils have high levels of unsaturated fatty acids, whereas animal fats have large amounts of saturated fatty acids. The commonly used fats are those from lard (Lee et al., 2002; Lu et al., 2007; Shin et al., 2012), lamb (Demirel et al., 2006; Taher et al., 2011), tallow (Chung et al., 2009; Nelson et al., 1996), and chicken (Shi et al., 2013). Typically, animal fats are cheap raw materials and using them is considered a waste management process. In addition, cooking oils are usually broken down after a period use and become unsuitable for further cooking as a result of increasing free fatty acid content (Marmesat et al., 2007). In the past, these waste products were used as an ingredient in animal feed, but they were later banned due to concerns regarding animal health. This waste cannot be discharged in sewers, as it causes blockages. Utilizing them for fuel can reduce biodiesel production costs and help in solving waste disposal problems. In addition, it is cheap feedstock (Phan and Phan, 2008). Nevertheless, these two alternatives, namely, animal fat and waste cooking oil, suffer from inconsistent supplies and logistical difficulties, and they cannot satisfy the existing global demand for diesel fuel (Taufiqurrahmi et al., 2011; Zhang et al., 2003).

6.2.4 MICROALGAE LIPIDS

Microalgae are unicellular photosynthetic organisms that convert inorganic carbons, such as CO_2, in the presence of light, water, and nutrients to algal biomass (Adamczak et al., 2009; Demirbas, 2010; Demirbas and Demirbas, 2011; Graham and Wilcox, 2000; Pokoo-Aikins et al., 2009; Vyas et al., 2010). They can grow either autotrophically or heterotrophically. The autotrophic growth is preferable as it fixes CO_2, which has a positive effect on the environment, and does not require any food source, such as glucose. The statistics showed that there are more than 40,000 species of algae, but only a limited number have been studied or have commercial significance (Harwood and Guschina, 2009; Peng et al., 2008), and the predominant group for biodiesel production is green microalgae.

The main components of any microalgae cell are protein, carbohydrates, and lipids, as shown in Table 6.3. In addition to the strain, the composition also depends on the growth medium, salinity, pH, temperature, and light intensity. This variation in composition allows microalgae to be used for different applications, ranging from food production to biofuels. Usually, microalgae are used as additives in animal food (Knauer and Southgate, 1999) and as nutritional supplements for human food (Becker, 2004), as with *Spirulina platensis*. They are also used in cosmetics as thickening agents and in agriculture as biofertilizers and soil conditioners (Chisti, 2007). Additionally, microalgae can be used to mitigate atmospheric CO_2.

TABLE 6.3

Chemical Composition of Various Microalgae Species (% Dry Weight)

Microalgae	Carbohydrates	Protein	Lipids	Reference
Chaetoceros muelleri	19.3	46.9	33.2	Voltolina et al., 2008
Isochrysis galbana	26.8	47.9	14.5	Natrah et al., 2007
Chaetoceros calcitrans	27.4	36.4	15.5	Natrah et al., 2007
Isochrysis sp.	12.9	50.8	20.7	Renaud et al., 2002
Prymnesiophyte (NT19)	8.4	41.3	14.7	Renaud et al., 2002
Rhodomonas (NT15)	6	57.2	12	Renaud et al., 2002
Cryptomonas (CRF101)	4.4	44.2	21.4	Renaud et al., 2002
Chaetoceros (CS256)	13.1	57.3	16.8	Renaud et al., 2002
Chlorella protothecoides[a]	10.6	52.6	14.6	Miao and Wu, 2004, 2006; Miao et al., 2004
Chlorella protothecoides[b]	15.4	10.3	55.2	Miao and Wu, 2004, 2006
Microcystis aeruginosa	11.6	30.8	12.5	Miao and Wu, 2004
Nannochloropsis sp.	29	10.7	60.7	Fábregas et al., 2004
Scenedesmus obliquus	15	50	9	Repka et al., 1998
Oscillatoria limnetica	50	44	5	Repka et al., 1998
Botryococcus braunii	18.9	17.8	61.4	Singh and Kumar, 1992
Botryococcus protuberans	16.8	14.2	52.2	Singh and Kumar, 1992

[a] Autotrophic cultivation.
[b] Heterotrophic cultivation.

Lipids are the cellular component used for biodiesel production. The content is usually between 20% to 50% of dry algae biomass weight, but in some cases may exceed 80% (Adamczak et al., 2009; Chisti, 2007; Spolaore et al., 2006; Vijayaraghavan and Hemanathan, 2009). Using lipids extracted from microalgae as a feedstock for biodiesel production was introduced by the National Renewable Energy Laboratory (NREL) research project (Sheehan et al., 1998). Due to their simple cell structure, they are widely used and have been generally accepted as a promising feedstock for biodiesel production. Microalgae of low lipid content (10 wt%) can still produce 4 times the amount of oil obtained from the best crop feedstock (palm). Microalgae can grow very fast by doubling their biomass in 24 h, and during their exponential growth phase they can double their biomass in about 3.5 h (Chisti, 2007; Meng et al., 2009; Patil et al., 2008; Vijayaraghavan and Hemanathan, 2009). In addition, several micro-algae strains can alter their metabolic pathways toward the accumulations of lipids in the form of triglycerides (Peng et al., 2008). Under stressed conditions, photo-synthesis activity decreases, but lipid synthesis is enhanced. Nitrogen stress is the most commonly reported factor that triggers lipid accumulation in green microalgae. Flow cytometry and Nile red (red phenoxazone dye) florescence and Fourier transform infrared spectroscopy (FTIR) are commonly used for rapid screening and detection of triglycerides and their enhancement in the cell (Satpati and Pal, 2014; Taher et al., 2014a). Figure 6.1 shows an example of lipids accumulation in *Scenedesmus* sp., due to nitrogen starvation, and it was monitored using Nile red staining.

FIGURE 6.1   **(See color insert.)** Fluorescence images of lipids accumulation in *Scenedesmus* sp. after (a) 1, (b) 11, (c) 14, (d) 20, and (e) 23 days of nitrogen starvation. (From Taher, H., S. Al-Zuhair, A. H. Al-Marzouqi, Y. Haik, and M. Farid, 2014, "Effective Extraction of Microalgae Lipids from Wet Biomass for Biodiesel Production," 66:159–167. With permission.)

Generally, microalgae lipids include neutral lipids, polar lipids, wax esters, sterols, and hydrocarbons, as well as phenyl derivatives (Naik et al., 2010). For biodiesel production, natural lipids, which are nonpolar, are always targeted. Typically, most of the lipids produced from microalgae strains that have been tested for biodiesel production have fatty acid constitutions similar to most common vegetable oils (Becker, 2004; Huang et al., 2012), as shown in Table 6.1.

The microalgae biodiesel production process consists of several steps, which are strain selection, biomass cultivation, harvesting, drying, cell disruption, lipid extraction, and, finally, biodiesel production. Each step has its own technology, challenges, and limitations; and for sustainable production, biomass production, recovery, and processing aspects have to be considered. Biodiesel production from microalgae lipids has not yet been commercialized despite the benefits of microalgae and the fact that they contain considerable numbers of natural lipids with a similar chemical composition to vegetable oils. This is mainly due to the high energy and costs of harvesting, drying, lipid extraction, and biodiesel production.

6.3   BIODIESEL PRODUCTION TECHNOLOGIES

The inventor of the diesel engine, Rudolf Diesel, was the first to use vegetable oils when he tested peanut oil in 1900. Though the oil could be used directly in the engine, it was problematic and had many shortcomings, such as high viscosity and low volatility, which caused poor atomization and led to incomplete combustion, operational problems, and carbon deposits (Akoh et al., 2007; Basha et al., 2009; Fukuda et al., 2001; Sharma et al., 2008). These problems can be overcome if the oil is mixed with conventional diesel in an appropriate ratio. However, this approach is still impractical for long-term use (Demirbas, 2009a,b; Rathore and Madras, 2007; Robles-Medina et al., 2009). Therefore, considerable efforts have been made to develop vegetable oil derivatives that have properties similar to those of petroleum diesel fuels. Dilution (blending), pyrolysis (thermal cracking), microemulsion, and transesterification are the most commonly accepted methods for minimizing the problems associated with the direct use of vegetable oils in compression engines (Fukuda et al., 2001; Helwani et al., 2009; Ma and Hanna, 1999).

Commercially, biodiesel is produced by transesterification, also known as alcoholysis, which is a chemical reaction between triglycerides and alcohols (Fan and

Burton, 2009; Helwani et al., 2009). A typical form of this reaction is shown in Chapter 2. The fatty acid chains present in the feedstock usually range from 12 to 22 carbon atoms. In these types of reactions, the triglycerides are transformed into straight-chain molecules by exchanging the alcohol from an ester with another alcohol in a process similar to hydrolysis, except that an alcohol is used instead of water. The final product is always similar in size to the molecules of the species present in the diesel fuel (Rajan and Senthilkumar, 2009; Sinha et al., 2008). Therefore, such products can be used in normal engines without any modification. By transesterification, oil molecular weight can be reduced to approximately one-third and the viscosity to about one-seventh, and the flash point and in some cases the volatility can be reduced as well (Demirbas, 2009b).

In the stoichiometry of transesterification reactions, each mole of triglyceride reacts with three moles of alcohol. However, a higher molar ratio of alcohol to oil is usually required to shift the reaction into biodiesel production. Methanol, ethanol, and propanol are the predominant alcohols. Methanol is the most commonly used, due to its low cost and industrial availability, which are essential for large-scale industrial production. On the other hand, ethanol has been gaining popularity in recent years due to its renewable possibilities and low toxicity sources compared to methanol. Extensive studies have been carried out for fatty acid ethyl ester production, and the properties of the biodiesels produced have been found to be similar to those of fatty acid methyl esters. However, the reaction rate was slower to a certain extent. Another major difference between using methanol and ethanol is that during ethanolysis, the glycerol–ester emulsion is more stable, resulting in a more difficult separation and downstream processing. Whereas when methanol is used, the emulsion can easily break down by ceasing mechanical stirring after the completion of the reaction (Encinar et al., 2007; Mendow et al., 2011; Zhou et al., 2003).

Typically, triglyceride and short-chain alcohols are immiscible, resulting in two phases with poor surface contact, which yields relatively slow reactions. However, these rates can be significantly increased by introducing a catalyst or carrying the reaction at a supercritical state (Fjerbaek et al., 2009; Lin et al., 2009). The catalytic reaction can be carried out using a catalyst, where by the end of the reaction, the biodiesel produced and the glycerol can be separated and purified to remove by-products and the catalyst. Catalyst selection depends on various factors. The most important of which is the free fatty acids (FFAs) and moisture contents of the feedstock.

6.3.1 ALKALI CATALYSTS

Alkali catalysts are commonly used in transesterification reactions because of their relatively low cost and ease of handling (Atadashi et al., 2012). In addition, transesterification reactions can be performed at low temperatures and pressures with a very high conversion yield, reaching 98% in a short time (Fukuda et al., 2001). Sodium hydroxide (NaOH), potassium hydroxide (KOH), and sodium methoxide (CH_3ONa) are the most common homogeneous alkali catalysts employed (Demirbas, 2009a,b; Helwani et al., 2009; Sharma et al., 2008). Many researchers have studied the use of these catalysts in transesterification reactions with oils of different FFAs content, ranging from 5 wt% to 15 wt%, as listed in Table 6.4. However, the use of

TABLE 6.4

Research Using Homogeneous Chemical Catalyzed Transesterification of Different Feedstocks for Biodiesel Production

Catalyst	Feedstock	Alcohol	Temperature (°C)	Molar Ratio (Alcohol:Oil)	Reaction Time	% Yield	Reference
			Base Catalysts				
NaOH	Duck tallow	Methanol	65	6:01	3 h	62.3	Chung et al., 2009
	Treated waste cooking oil	Methanol	50	7:01	1 h	88.9	Meng et al., 2008
	Sunflower oil	Methanol	65	6:01	–	86.7	Vicente et al., 2004
	Neat canola oil	Methanol	70	6:01	15 min	93.5	Leung and Guo, 2006
	Treated used frying oil	Methanol	60	7:01	20 min	88.8	Leung and Guo, 2006
	Soybean oil	Methanol	45	6:01	20 min	100	Ji et al., 2006
	Sunflower oil	Methanol	60	6:01	1.5 h	97.1	Rashid et al., 2008
	Sunflower oil	Ethanol	50	12:01	10 min	99	Marjanovic et al., 2010
	Soybean oil	Ethanol	70	12:01	1 h	97.2	Kucek et al., 2007
	Waste cooking oil	Methanol	50	6:01	1.5 h	86	Meng et al., 2008
KOH	Duck tallow	Methanol	65	6:01	3 h	79.7	Chung et al., 2009
	Karanja oil	Methanol	65	6:01	3 h	97–98	Meher, Vidya et al., 2006
	Pongamia pinnata oil	Methanol	45	10:01	1.5 h	83	Karmee and Chadha, 2005
KOH	Soybean oil	Ethanol	70	12:01	–	95.6	Kucek et al., 2007
	Pongamia pinnata oil	Methanol	60	10:01	1.5 h	92	Karmee and Chadha, 2005
	Mixture of castor and soybean	Ethanol	–	19:02	4 h	97	Barbosa et al., 2010
	Sunflower oil	Methanol	65	6:01	–	91.6	Vicente et al., 2004
	Fish oil	Ethanol	60	6:01	30 min	98	Armenta et al., 2007
	Waste cooking oil	Methanol	65	6:01	1 h	98.16	Refaat et al., 2008

(Continued)

TABLE 6.4 (CONTINUED)

Research Using Homogeneous Chemical Catalyzed Transesterification of Different Feedstocks for Biodiesel Production

Catalyst	Feedstock	Alcohol	Temperature (°C)	Molar Ratio (Alcohol:Oil)	Reaction Time	% Yield	Reference
CH_3NaO	Duck tallow	Methanol	65	6:01	3 h	79.3	Chung et al., 2009
	Sunflower oil	Methanol	65	6:01	–	99.3	Vicente et al., 2004
	Castor oil	Ethanol	80	6:01	3 h	80	Meneghetti et al., 2006
	Coriander seed oil	Methanol	60	6:01	1.5 h	94	Moser and Vaughn, 2010
Acid Catalysts							
H_2SO_4	Waste cooking oil	Methanol	95	20:01	20 h	90	Ji et al., 2006
	Used sunflower oil	Methanol	65	30:01:00	69 h	99	Freedman et al., 1984
	Waste frying oil	Methanol	70	245:01:00	4 h	99	Zheng et al., 2006
H_2SO_4	Rice bran oil	Methanol	60	5:01	12	99	Zullaikah et al., 2005
Two Steps: Acid Catalysis Followed by Alkali Catalysis							
H_2SO_4	Karanja	Methanol	60	6:01	1 h	FFAs	Sharma et al., 2009
KOH			60	8:01	1 h	96	
HCL	Fish frying oil	Methanol	60	6:01	1	FFAs	Fadhil et al., 2012
KOH			60	6:01	1	94	
Ferric sulfate	Waste cooking oil	Methanol	95	10:01	2 h	FFAs	Ji et al., 2006
KOH			65	6:01	1 h	97	
Ferric sulfate	Waste cooking oil	Methanol	100	9:01	2 h	FFAs	Patil et al., 2010
KOH			100	7.5:1	1 h	96	
H_2SO_4	*Dinoflagellate* oil	Methanol	65	NA	2–3 h	FFAs	Chen et al., 2012
KOH			65	NA	1 H	90.1	

alkali catalysts is not practical with feedstock of relatively high FFAs, due to soap formation, which reduces the quality of the final product and adds difficulty to the separation step. The production process starts with mixing the catalyst with alcohol to produce the alkoxide solution. This is then charged with the oil in a closed vessel. The reaction mixture is then heated to just near the boiling point of the alcohol to speed up the reaction. After the reaction is complete, the products are separated and excess alcohol is recovered. The two products, biodiesel and glycerol, are usually separated in a settling vessel. As glycerol is much denser than biodiesel, biodiesel floats to the top, and both phases would have a significant amount of excess alcohol that needs to be recovered at the end.

As mentioned earlier, the quality of the feedstock has a significant effect on the reaction yield, where FFAs and moisture contents are key parameters in defining feedstock feasibility. For example, the alkali-catalyzed production of biodiesel from duck tallow at 65°C and a 6:1 methanol-to-oil molar ratio resulted in a 62% conversion (Chung et al., 2009), whereas sunflower resulted in an 86% conversion using the same catalyst under the same conditions (Vicente et al., 2004).

Although alkali-catalyzed biodiesel is commercially viable with high yields, this type of reaction has several limitations. It is more suited to feedstocks of low FFA content, usually below 0.5 wt% (Ma et al., 1998; Sivasamy et al., 2009). The presence of FFAs promotes soap formation, which consumes and hinders the catalyst, lowers the production yield, and more importantly causes difficulties in the downstream separation of the esters formed from the glycerol due to the formation of complex emulsions (Al-Zuhair, 2007; Ma and Hanna 1999). Therefore, refined feedstocks are required. These are usually expensive. The refining is usually carried out by distillation at high temperature under reduced pressure. This may cause degradation of temperature-sensitive compounds and add costs to the overall process. Another drawback of alkali-catalyzed processes is the requirement to purify the products from the catalyst. Such a process needs large quantities of water, which in turn, requires further treatment. In addition, some oils contain phospholipids that require degumming via hydration prior to use (Carelli et al., 1997). Having said that, not all phospholipids are hydratable, and citric or phosphoric acids are used to convert them into hydratable gum (Verleyen et al., 2002).

To overcome the high cost of wastewater treatment processes and the cost of the recovery of the catalyst and product, heterogeneous catalysts were developed. These catalysts can be easily separated and reused after simple filtration (Granados et al., 2007; Sharma et al., 2008). The processes in this case are environmentally benign with minimal wastewater treatment required. Alkaline earth oxides (Ebiura et al., 2005), zeolites (Peterson and Scarrah, 1984), calcined hydrotalcites (Silva et al., 2010; Xie et al., 2006), and magnesium and calcium oxides (Dossin et al., 2006; Granados et al., 2007) have been extensively tested. Among these, calcium oxide (CaO) has received considerable attention due to its relatively low cost, high core strength, and low level of methanol (Zabeti et al., 2009). Liu et al. (2008) investigated the use of CaO in the transesterification of soybean oils to biodiesel, and tested the effects on the molar ratio of methanol to oil, the reaction temperature, and water content. It was found that increasing the reaction temperature from 50°C to 65°C resulted in an increased yield by a factor of 2.4 at a 6:1 molar ratio. However, a further increase resulted in a significant drop in the yield, due to methanol vaporization. The yield for biodiesel at 65°C also increased from 61% to 97% with an increase in the methanol-to-oil molar ratio from 3:1 to 12:1. It was also found that CaO maintained its activity for 20 cycles without a significant drop (Helwani et al., 2009; Lam et al., 2010). Table 6.5 shows a summary of works carried out using heterogeneous alkali catalysts. The most obvious problem of using these heterogeneous base catalysts is a slower reaction rate when compared to homogeneous catalysts. This is because the reaction mixture is in three separate phase systems that cause diffusion problems and require higher reaction temperatures, higher catalyst loading, and a higher alcohol-to-oil molar ratio. Although the use of heterogeneous catalysts minimizes environmental issues, catalyst leaching during the reaction is inevitable. In addition, the high cost of purified feedstock remains the main problem facing alkali-catalyzed processes (Helwani et al., 2009; Lam et al., 2010).

6.3.2 Acid Catalysts

Transesterification of oils can also be catalyzed using acid instead of alkaline catalysts. Acid-catalyzed transesterification is more suitable with unpurified feedstocks with a high FFA content. These are also much cheaper (Al-Zuhair, 2007; Al-Zuhair et al., 2007; Zhang et al., 2003). The most commonly used catalysts are strong acids; such as sulfuric, sulfonic, phosphoric, and hydrochloric acids (Fukuda et al., 2001; Ma et al., 1999; Meher, Kulkarni et al., 2006). However, these processes are not as popular as alkaline-catalyzed processes, mainly because of corrosion, which necessitates specific equipment, which in turn increases the overall production costs. In addition, acid-catalyzed reactions are too slow. They are reported to be 4000 times slower than their alkaline counterparts (Akoh et al., 2007; Al-Zuhair, 2007; Marchetti et al., 2007). Moreover, acid-catalyzed reactions are usually performed at high alcohol-to-oil molar ratios, reaching as much as 30:1 with high catalyst concentrations (Akoh et al., 2007; Helwani et al., 2009). In some case, the alcohol-to-oil molar ratio reached up to 245 (Zheng et al., 2006). Although biodiesel production costs can be reduced by using low cost feedstocks with high FFA content, acid-catalyzed transesterifications of these low cost feedstocks require high catalyst concentrations, and the neutralization and removal of the catalyst from the product. Table 6.4 also shows a summary of work using acid catalysts in the biodiesel production process.

A two-step transesterification process has been suggested (Liu et al., 2006), where an acid-catalyzed esterification is used to lower the FFA content, followed by an alkali-catalyzed transesterification after the separation of the acid catalyst and the production of water (Canakci and Van Gerpen, 2001). The separation of water is an important step prior to utilizing an alkali-catalyzed process if you wish to avoid soap formation. Several studies have been carried out using this two-step approach with yields reaching as high as 98% (Berchmans and Hirata, 2008; Canakci and Van Gerpen, 2001; Chen et al., 2012; Ghadge and Raheman, 2005; Sathya and Manivannan, 2013), as given in Table 6.4.

TABLE 6.5

Research Using Conventional Heterogeneous Chemical-Catalyzed Reactions of Different Feedstocks for Biodiesel Production

Catalyst	Feedstock	Alcohol	Temperature (°C)	Molar Ratio (Alcohol:Oil)	Reaction Time (h)	% Yield	Reference
			Heterogeneous Base Catalysts				
CaO	Soybean oil	Methanol	60–65	12:01	1	66	Kouzu et al., 2008
	Refined soybean oil	Methanol	60–65	12:01	1	93	Kouzu et al., 2008
	Refined soybean oil	Methanol	65	12:01	3	95	Liu et al., 2008; Meher, Kulkarni et al., 2006
Li/CaO	Karanja oil	Methanol	65	12:01	8	95	Meher, Kulkarni et al., 2006
K_3PO_4	Waste cooking oil	Methanol	60	18:01	2	97.3	Guan et al., 2009
			Heterogeneous Acid Catalysts				
WO_3/ZrO_2	Unspecified vegetable oil	Methanol	75	19.4:1	20	85.1	Park et al., 2008
ZrHPW	Waste cooking oil	Methanol	65	20:01	8	98.9	Zhang et al., 2009
ZS/Si	Waste cooking oil	Methanol	200	18:01	5	98	Jacobson et al., 2008
$SO_4^{2-}/TiO_2–SiO_2$	Waste cooking oil	Methanol	500	9:01	4	90	Peng et al., 2008
$SO_4^{2-}/SnO_2–SiO_2$	Waste cooking oil	Methanol	150	15:01	3	92.3	Lam et al., 2009

6.3.3 NONCATALYTIC SUPERCRITICAL TRANSESTERIFICATION

Although catalysts reduce reaction time, their presence complicates final product purification. This results in increased overall production costs. To avoid this, alcohols are used at above their critical temperature and pressure (Demirbas, 2002, 2007a, 2009a). Supercritical transesterification is a one-step conversion process performed when the alcohol is in a supercritical state and in the absence of a catalyst. Only reactants are added and heated or pressurized. Table 6.6 shows a summary of studies carried out using supercritical alcohols. Methanol and ethanol are the most common alcohols used. Typical operating conditions are 250°C–400°C and 150–400 bar of pressure. Usually such processes result in a much faster (within 5–15 minutes), simpler, and more environmentally friendly product separation when compared to catalyzed processes. Pretreating the feedstock, purifying the product, and treating the wastewater are not required. Saka and Kusdiana (2001) investigated biodiesel production from rapeseed oil at above supercritical conditions, and a high conversion rate of 95% was achieved with this method. Demirbas (2002) also studied the transesterification of cotton, hazelnut kernels, poppy, rapeseed, safflower seeds, and sunflower seeds, and reported that increasing the reaction temperature to above supercritical had a positive influence on the conversion. Compared to catalytic reactions, supercritical reactions are fast and can achieve high conversions in a very short time.

Although supercritical alcohol reactions seem promising, as they solve most of the existing problems, there has been a lot of debate on the efficiency of the reaction in terms of energy utilization and also safety issue due to the high pressure and temperature employed in this technique (Rajan and Senthilkumar, 2009; Sinha et al., 2008). Hence, there are issues and challenges that need to be addressed before this technology can play a major role in biodiesel production.

Generally, the three transesterification processes—alkaline- and acid-catalyzed processes and supercritical noncatalytic processes—have several drawbacks. The use of enzymes, especially lipase, can overcome these problems, and this presents a more environmentally friendly alternative at lower temperatures (Al-Zuhair, 2007; Haas and Foglia, 2005).

6.3.4 BIOCATALYSTS

There is also some value to using biocatalysts in biodiesel production. Regardless of their sources, lipases can hydrolyze triglycerides (Marchetti et al., 2007) in mild operating conditions. Thus, they require less energy and have the ability to work with triglycerides from different origins. Unlike chemical catalysts, lipases do not form soaps and can simultaneously catalyze both triglycerides transesterification and FFA esterification. This diversity allows them to be used in various applications. In addition, enzymatic transesterification generates a product of high purity with very few downstream operations (Fukuda et al., 2001).

The lipase-catalyzed production of biodiesel was first reported in 1990 by Mittelbach, who considered different lipases to catalyze the alcoholysis of sunflowers with short-chain alcohols in ether. For optimal production, the lipases should be

TABLE 6.6
Noncatalytic Vegetable Oils Transesterification Using Supercritical Alcohols for Biodiesel Production

Feedstock	Alcohol	Molar Ratio (Alcohol:Oil)	Reaction Conditions Temperature (°C)	Pressure (bar)	Time (min)	% Yield	Reference
Palm	Methanol	40:1	360	–	15	81.1	Tan et al., 1988
Krating	Methanol	40:1	260	160	10	90.4	Sanniang et al., 2014
Jatropha	Methanol	40:1	320	150	5	84.6	Sanniang et al., 2014
Lard	Methanol	45:1	335	200	15	89.9	Shin et al., 2012
Macauba	Methanol	30:1	325	150	–	78.5	Navarro-Diaz et al., 2014
Soybean	Ethanol	40:1	350	200	–	77.5	Vieitez et al., 2008
Palm	Ethanol	40:1	360	–	150	79.5	Tan et al., 2010
Rapeseed	Ethanol	42:1	300	250	12	100	Demirbas, 2003
Rapeseed	Ethanol	42:1	300	400	20	80	Demirbas, 2003
Cottonseed	Ethanol	41:1	250	–	8	85	Demirbas, 2003
Linseed	Ethanol	41:1	250	–	8	85	Demirbas, 2009a
Sunflower	Ethanol	40:1	300	200	40	100	Madras et al., 2004

nonspecific; and the triglycerides, diglycerides, and monoglycerides are all attacked by the lipase. As such, lipase from *Candida antarctica*, which is nonspecific, is commonly used (Fjerbaek et al., 2009). This has been tested with a variety of substrates and alkali acceptors and shows promising results (Nelson et al., 1996). Other commonly used lipases are those from *Pseudomonas flseudomona*, *P. cepacia*, *C. rugosa*, *Rizhomucor miehei*, and *Thermomyces lanuginosa*.

6.4   LIPASE-CATALYZED TRANSESTERIFICATION: TECHNICAL CHALLENGES

Despite the advantages of lipases over chemical catalysts, biodiesel production using lipases has not yet been commercialized. To achieve this, several technical and economic challenges need to be overcome. The most important is the high cost of the enzyme, compared to chemical catalysts, and the inhibition by alcohol. Therefore, selection of the lipase and the optimization of reaction parameters are important aspects that have to be considered to obtain increased and purified biodiesel yields.

6.4.1   Cost

The most important challenge in enzymatic biodiesel production is the high cost of the lipase. Therefore, immobilization was considered in order to reduce overall biodiesel production costs. Lu et al. (2007) transesterified lard using immobilized *Candida* sp. 99–125 and found that the enzyme was reusable over seven repeated cycles (for 180 hours) with no significant decrease in activity. Also the production yield was higher than 80%. Modi et al. (2007) found a similar stability when ethyl acetate was used instead of alcohol; they used Novozym®435. The immobilized enzyme was reused for 12 cycles without any loss in the activity. On the other hand, Shimada et al. (1999) reported more than 95% conversion even after 50 cycles (100 days) of the reaction. Table 6.7 shows a summary of different lipases tested with different feedstocks, conditions and immobilization methods.

Although immobilization can reduce the overall production cost, the overall cost is still much higher than that of chemical-catalyzed processes. Jegannathan et al. (2011) determined the economic feasibility of producing 1000 tons of biodiesel from palm oil using alkali and lipase catalysts in both soluble and immobilized forms. The alkali-catalyzed process cost was found to be lowest followed by the immobilized form of the lipase. However, in this research it was assumed that lipase could be reused only five times. If reusability is increased, the lipase processes becomes more feasible.

6.4.2   Inhibition by Methanol

Although short-chain alcohols, specifically methanol, are preferable in conventional transesterification reactions, they negatively affect the activity of lipase if used in excessive amounts. This is because short-chain alcohols have low solubility in oils, resulting in their presence as a separate phase. The affinity of the alcohol to water is high, which results in a stripping of the microlayer of essential water surrounding

TABLE 6.7
Research for Biodiesel Production from Different Lipases Immobilized by Different Methods

Lipase	Method	Carrier	Fat/Oil	Acyl Acceptors	% Yield	Reference
C. antarctica	Adsorption	Silica Gel	Soybean	Methanol	94	Wang et al., 2006
	Adsorption	Anion resin	Palm kernel	Ethanol	63	Oliveira and Vladimir Oliveira, 2001
	Adsorption	Acrylic resin	Soybean	Methyl acetate	92	Du et al., 2004
	Adsorption	Acrylic resin	Soybean	Methanol	92.8	Modi et al., 2006
	Cross linking	Glutaraldehyde	Madhuca	Ethanol	92	Kumari et al., 2007
P. cepacia	Adsorption	Celite	Jatropha	Ethanol	98	Shah and Gupta, 2007
	Entrapment	Hydrophobic sol-gel	Soybean	Methanol	56	Noureddini et al., 2005
	Entrapment	Phyllosilicate sol-gel	Tallow greases	Ethanol	94	Hsu et al., 2001
	Encapsulation	Burkholderia cepacia	Sunflower	Methyl acetate	64	Orçaire et al., 2006
P. fluorescens	Adsorption	Polypropylene EP 100	Sunflower	Methanol	91	Soumanou and Bornscheuer, 2003
	Adsorption	Polypropylene MP1004	Soybean	Methanol	96	Salis et al., 2008
	Entrapment	Sodium alginate	Jatropha	Methanol	72	Devanesan et al., 2007

the lipase, which is required to maintain lipase conformation and catalytic activity (Du et al., 2004; Fjerbaek et al., 2009; Li et al., 2006; Zheng et al., 2009). In addition, due to the high polarity of methanol, the excess binds to active sites on the enzyme and prevents the substrate from reaching them. Moreover, high molar ratios of alcohol-to-triglycerides increase glycerol solubility and affect its separation. Therefore, optimization of the molar ratio is a big challenge and has to be carefully considered. The negative effect of methanol on the reaction has been confirmed by many studies that have been done on enzymatic biodiesel production. Santambrogio et al. (2013) studied the effects of methanol on the activity and conformation of *Burkholderia glumae* lipase, and found that enzyme stability was negatively affected and caused protein unfolding and conformational changes. A similar observation was reported with *Rhizopus oryzae* (Chen et al., 2006). As in the case of methanol, the stepwise addition of ethanol resulted in a 60% ethyl ester yield compared to only 16.9% when it was added in a single step at the beginning of the reaction (Bernardes et al., 2007). Lee et al. (2002) reported a higher conversion rate of 74% using methanol as compared to 43% when ethanol was used in a three-step manner.

6.4.2.1   Different Acyl Acceptors

The use of other acyl acceptors has been suggested as a way to eliminate the inhibition caused by methanol. Among them, methyl and ethyl acetates were the most commonly used acceptors for interesterification of different oils and fats to produce biodiesel. Du et al. (2004) performed a comparative study on Novozym®435 transesterification of soybean oil with methanol and interesterification of the same oil with methyl acetate for biodiesel production. The high methanol concentration showed a negative effect on enzymatic activity, whereas methyl acetate did not have any visible negative effects. Xu et al. (2005) tested methyl acetate with Novozym®435 by using soybean oil at 40°C, and similar results were found. Modi et al. (2006, 2007) tested ethyl acetate and propan-2-ol when converting oils from jatropha, karanja, and sunflowers to biodiesel using Novozym®435 at 50°C. They found that the enzyme showed higher stability when used with both acceptors compared to use with ethanol and methanol. Despite methyl acetate's clear advantages for enzyme activity, it has a much slower reaction rate (Du et al., 2004). Tables 6.7 and 6.8 list examples of studies using ethyl and methyl acetates in lipase-catalyzed biodiesel production.

6.4.2.2   Lipase Pretreatments

Numerous studies have reported that pretreatment of an immobilized lipase affects its activity and stability. Samukawa et al. (2000) studied the effect of Novozym®435 preincubation in methyl oleate on soybean oil transesterification with methanol. Immobilized lipase was incubated in methyl oleate for 0.5 h and then in the oil for 12 h. The use of an incubated enzyme resulted in a 20% higher yield in 1 h, compared to only 13.6% achieved by the nonincubated enzyme. The stability of immobilized lipase was also enhanced by incubation in *tert*-butanol. The high catalytic activity and stability of the incubated immobilized lipase have been confirmed by several researchers (Li et al., 2006; Royon et al., 2007). The positive effect might be due to *tert*-butanol's ability to shield the enzyme from being inhibited by methanol or by immersion in *tert*-butanol.

TABLE 6.8

Biodiesel Production Studies Using Different Immobilized Lipases and Transesterification Methods

Lipase	Fat/Oil	Alcohol	Solvent	Temperature (°C)	Molar Ratio (Alcohol:Oil)	Time (h)	% Yield	Reference
Novozym®435	Cotton seed	Methanol	tert-Butanol	50	4:01	24	97	Royon et al., 2007
	Cottonseed	Methanol	tert-Butanol	51	5:01	25	95	Royon et al., 2007
	Jatropha	Ethyl acetate	–	50	11:01	12	91.3	Modi et al., 2007
	Soybean	Methanol	tert-Butanol	50	4:01	12	65.8	Ha et al., 2007
	Soybean	Methyl acetate	–	40	12:01	14	92	Xu et al., 2003
	Soybean	Methyl acetate	–	40	12:01	14	92	Du et al., 2004
	Palm kern	Ethanol	n-Hexane	40	10:01	4	58.3	Oliveira and Vladimir Oliveira, 2001
	Palm kern	Ethanol	SC-CO$_2$	40	10:01	4	63	Oliveira and Vladimir Oliveira, 2001
	Sunflower	Methanol	SC-CO$_2$	45	5:01	6	23	Madras et al., 2004
	Soybean and rapeseed	Methanol	–	30	–	–	33.1	Shimada et al., 1999
	Sunflower	Ethanol	SC-CO$_2$	45	5:01	6	27	Madras et al., 2004
	Soybean	Methanol	–	30	–	3.5	97	Samukawa et al., 2000
P. fluorescens	Sunflower	1-propanol	–	60	3:01	20	91	Iso et al., 2001
	Soybean	Methanol	–	35	6:01	90	80	Kaieda et al., 2001
	Sunflower	Methanol	1,4-Dioxane	50	3:01	80	70	Iso et al., 2001
	Jatropha	Methanol	n-Hexane	40	4:01	48	71	Devanesan et al., 2007

(Continued)

TABLE 6.8 (CONTINUED)

Biodiesel Production Studies Using Different Immobilized Lipases and Transesterification Methods

Lipase	Fat/Oil	Alcohol	Solvent	Temperature (°C)	Molar Ratio (Alcohol:Oil)	Time (h)	% Yield	Reference
P. cepacia	Tallow + Grease	Ethanol	–	50	4:01	20	94	Hsu et al., 2001
	Soybean	Methanol	–	35	6:01	90	100	Kaieda et al., 2001
	Jatropha	Ethanol	–	50	4:01	12	98	Shah et al., 2004
	Sunflower	2-butanol	–	40	3:01	6	100	Salis et al., 2005
	Madhuca	Ethanol	–	40	4:01	2.5	92	Kumari et al., 2007
	Soybean	Methanol	–	35	7.5:1	1	67	Noureddini et al., 2005
	Soybean	Ethanol	–	35	7.5:1	1	65	Noureddini et al., 2005
	Mahua	Ethanol	–	40	4:01	6	96	Kumari et al., 2007
Lipozyme IM	Palm kern	Ethanol	n-Hexane	40	3:01	4	77.5	Oliveira and Vladimir Oliveira, 2001
	Palm kern	Ethanol	SC-CO$_2$	40	6.5:1	4	26.4	Oliveira and Vladimir Oliveira, 2001
	Tallow	Methanol	–	45	3:01	5	19.4	Nelson et al., 1996
	Tallow	Ethanol	–	45	3:01	5	65.5	Nelson et al., 1996
	Tallow	Ethanol	n-Hexane	45	3:01	5	98.3	Nelson et al., 1996
Lipozyme IM	Soybean	Methanol	–	40	3:01	12	98	Xu et al., 2004
	Soybean	Ethanol	n-Hexane	45	3:01	5	97.4	Nelson et al., 1996
Candidia sp. 99–125	Microalgae	Methanol	n-Hexane	38	3:01	12	98.15	Li et al., 2007
Lipozyme TL IM + Novozyme®435	Rapeseed	Methanol	tert-Butanol	35	4:01	12	95	Li et al., 2006

6.4.2.3 Reaction Engineering

Enzyme inhibition also can be overcome through medium engineering, specifically improving the medium's polarity by adding an inert solvent. Generally, introducing an organic solvent to synthetic reactions increases the solubility of substrates with a similar polarity, thus increasing the reaction rate and reducing water activity, all of which diminishes hydrolysis during the transesterification reaction (Radzi et al., 2005). Moreover, conducting reactions in an organic solvent medium has other advantages compared to solvent-free systems. These include reducing the reaction medium's viscosity, which allows for higher diffusion with fewer mass transfer limitations. In addition, there is better stability, and easy recovery and reuse of the immobilized enzyme from the reaction mixture (Klibanov, 2001; Zaks and Klibanov, 1984).

The selection of a suitable solvent is critical, as solvents may also strip essential hydration shells from the enzyme. This is similar to short-chain alcohols. Although numerous organic solvents can be used, several aspects have to be considered when choosing a suitable solvent. These include solvent compatibility, inertness, low density, toxicity, and flammability (Adamczak and Krishna, 2004). Lipases have been tested in various organic solvents such as *tert*-butanol and *n*-hexane (Eltaweel et al., 2005; Hernández-Rodríguez et al., 2009; Li et al., 2006).

The importance of using a hydrophobic solvent has been well addressed in the literature. Transesterification of cottonseed oil using Novozym®435 was carried out in a *tert*-butanol solvent medium and compared with a reaction in a solvent-free medium under the same conditions (Köse et al., 2002; Royon et al., 2007). It was found that by carrying out the reaction in a *tert*-butanol solvent, the yield obtained after 24 h was 97%, compared to only 92% in a solvent-free system. Using 80% *tert*-butanol also improved biodiesel yields from soybean oil deodorizer distillate using a 4% Novozym®435 from 80% to 84% in a solvent-free reaction (Ji et al., 2006). A more significant effect was observed when the *n*-hexane solvent was used in the transesterification of tallow fats with a 3:1 molar ratio to methanol and when using *Mucor miehei* lipase (Nelson et al., 1996). The yield obtained after 5 h was 94.8% in *n*-hexane, compared to only 19.4% obtained in a solvent-free system. When ethanol was used instead of methanol, a 65.5% yield was achieved in a solvent-free system and 98.0% in a *n*-hexane solvent in the first 5 h of the reaction. The higher yield obtained in a solvent-free system when ethanol was used, as compared to methanol, is mainly due to the lower inhibitory effect of ethanol.

Despite their positive effects on enzyme stability, excessive use of organic solvents decreases the yield, which is mainly due to the dilution of the substrates (Ji et al., 2006). Samukawa et al. (2000) tested the use of different solvents on Lipozyme TL activity, and it was found that the yield increased along with the hydrophobicity of the solvent. In contrast, hydrophilic solvents are much less efficient (Doukyu and Ogino, 2010; Klibanov, 1997). For example, the use of acetone ($\log P = -0.23$) showed a less than 20% yield compared to 80% when *n*-hexane was used. The negative effect of hydrophobicity is due to the solvent's interaction with the essential water layer surrounding the lipase molecule (Iso et al., 2001), which results in unfavorable conformational changes in the enzyme structure and reduces activity.

On the other hand, from an environmental point of view, the use of organic solvents should be minimized because of their harmful environmental impacts. In addition, they are usually expensive and require separation from the product. The need for high yields with easy separation to produce products of higher purity while using environmentally friendly processes has led to a search for new technologies and new solvents. Any alternative to organic solvents should dissolve reaction substrates and reduce the excess alcohol inhibition effect and, at the same time, avoid a difficult separation of the solvent. In this regard, supercritical fluids (SCFs) have been put forward and have shown potential (Romero et al., 2005). Further discussion on the use of SCFs, as a reaction medium, is covered in Section 6.5.

6.4.3 INHIBITION BY GLYCEROL

In addition to methanol inhibition, another inhibition affects lipase-catalyzed reactions. This is due to the hydrophilic nature of reaction by-product, glycerol (Caballero et al., 2009; Jegannathan et al., 2009; Noureddini et al., 2005; Robles-Medina et al., 2009; Sharma et al., 2008). Glycerol is insoluble in oils and adsorbs easily on immobilized lipase molecules. As such, it prevents the hydrophobic substrate from reaching the active site of the enzyme (Szczesna Antczak et al., 2009). In addition, glycerol drives the reaction equilibrium in a reverse reaction and increases the viscosity of the reaction medium. This, in turn, increases the time of transesterification contact. To minimize its accumulation, continuous removal from the reaction mixture is required. This can be achieved by running the reaction through a silica bed that adsorbs glycerol and removes it from the reaction medium (Yori et al., 2007). In addition, silica gels can also adsorb methanol and their addition to the reaction system decreases the inhibitory effect of methanol while increasing biodiesel production (Wang et al., 2006). In practice, although this method reduces inhibition to some extent, it complicates the system, especially in large-scale production. Dialysis using an ultrafiltration flat sheet membrane has also been suggested in order to remove the glycerol in continuous biodiesel production.

The use of acetates, rather than short-chain alcohols that do not produce glycerol, has also been suggested. When acetates are used, triacetin, which does not have the same negative effect on biodiesel yields as glycerol, is formed (Xu et al., 2003). However, triacetin separation is more complex than glycerol (Ruzich and Bassi, 2010). Moreover, organic solvents such as *tert*-butanol remain a preferred solution. The solvent dissolves glycerol and reduces its negative effect (Royon et al., 2007).

6.4.4 EFFECT OF TEMPERATURE

Like all chemically catalyzed reactions, reaction rate constants, and hence the reaction rate, increase with temperature. However, this is not always correct with enzyme-catalyzed reactions, due to protein denaturation at high temperatures. This results in the substrate no longer fitting the enzyme's active sites. Therefore, a rapid decline in the reaction rate and conversion yield is usually observed at high temperatures. In addition, the presence of the inactive denatured particles at the lipid interface blocks the active enzyme from being utilized, which results in a further drop in the yield. This trend was observed in studies that investigated the effect of temperature on the production of biodiesel using lipase (Köse et al., 2002; Taher et al., 2011).

The optimum temperature when using Novozym®435 was found to be 45°C to 55°C (Chen et al., 2011; Shaw et al., 2008; Taher et al., 2011). However, a lower optimum temperature of 38°C was observed by Chang et al. (2005) using canola oil with a 42.3% loading, 7.2% water, and a 3.5 molar ratio. This might be due to a high water concentration acting as a competitive inhibitor to transesterification at high temperatures.

6.5 LIPASE-CATALYZED REACTIONS WITH SUPERCRITICAL FLUIDS

In biodiesel production, methanol and lipid reaction products are immiscible and form two phases at room conditions. This results in low reaction efficiency and lipase deactivation. Hydrophobic solvents can minimize this effect, however, they are toxic and require a separation unit, which further increases the overall production cost. Supercritical carbon dioxide ($SC\text{-}CO_2$) has frequently been used to replace organic solvents in various chemical processes. Due to its properties—including the easy and complete removal of the solvent, an ability to manipulate the physical properties of the solvent by simply changing the pressure or temperature, nontoxicity, nonflammability, and enhancement of substrate mass transfer properties—it was suggested as a green solvent in biocatalyst reactions. A chemical feature of $SC\text{-}CO_2$ is its low critical temperature (below the denaturation temperature of lipase). This feature combines the good solubility of nonpolar compounds, such as lipids, and makes $SC\text{-}CO_2$ the perfect medium for biodiesel production.

Although high pressures are involved in such processes, pressures up to 300 bar were found to have a minimal effect on enzyme activity and stability in esterification reactions (Novak et al., 2003). Giessauf and Gamse (2000) reported an increase in pancreatic lipase activity when exposed to $SC\text{-}CO_2$ at 150 bar for 24 h. No loss in activity was reported when stored in this condition for 6 months.

As mentioned earlier, the biodiesel reaction mixture forms a two-phase system. However, in the presence of CO_2, at conditions above the critical value of CO_2, the reaction mixture dissolves in a single phase. This was observed when the behavior of a corn oil–methanol reaction mixture was visualized at different CO_2 pressures in 10 ml sealed view cells (Ciftci and Temelli, 2011). The same was found in a microalgae lipids–methanol mixture (Taher, 2014), as shown in Figure 6.2. At atmospheric pressure, the two separate phases were clearly shown (Figure 6.2a). These two phases exist up to a pressure of 60 bar, which is still below the critical pressure of CO_2 (Figure 6.2b). However, if the pressure increases to above the critical value of CO_2, then the reaction mixture dissolves in a single phase. The boundary between the two phases disappears and the mixture forms a single phase (Figure 6.2c,d)

The focus on lipase-catalyzed reaction in $SC\text{-}CO_2$ is due to enhanced reaction rates (Lee et al., 2013). In biodiesel production systems, $SC\text{-}CO_2$ offers easy product separation from the reaction mixture by selectively dissolving the biodiesel, due to its high solubility compared to the glycerol by-product (Rodrigues et al., 2011). In a continuous system, the product is continuously removed from the reaction system, which can then be easily separated from $SC\text{-}CO_2$ by simple depressurization.

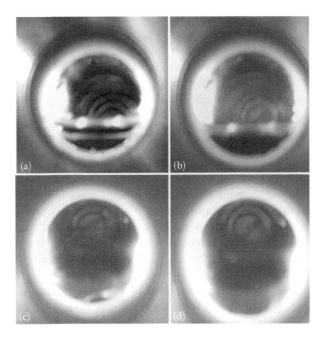

FIGURE 6.2   (See color insert.) Mixture of microalgae lipids and methanol in 10 ml view cells at 50°C; a 5:1 molar ratio; and (a) atmospheric pressures, (b) 60 bar, (c) 150 bar, or (d) 200 bar.

Promising results have been reported on the use of lipase with $SC\text{-}CO_2$ in the production of biodiesel from vegetable oils and animal fats. Kumar et al. (2004) esterified palmitic acid with ethanol at a temperature range of 35°C to 70°C in the presence of three different lipases under $SC\text{-}CO_2$ conditions. Their results showed that Novozym®435 was the best catalyst with a yield of 74%. Romero et al. (2005) made similar observations on the esterification of isoamyl alcohol in $SC\text{-}CO_2$ and n-hexane, where these showed higher reaction rates in $SC\text{-}CO_2$. Higher yields were also reported using $SC\text{-}CO_2$, as compared to n-hexane and solvent-free systems (Laudani et al., 2007) using Lipozyme RM IM.

Although $SC\text{-}CO_2$ has been used as a reaction medium for the enzyme esterification of FFAs, limited work has been done on transesterification. Oliveira and Vladimir Oliveira (2001) compared the enzymatic alcoholysis of palm kernel oil using n-hexane and $SC\text{-}CO_2$ systems. They discovered the highest conversion (63.2%) using Novozym®435. Rathore and Madras (2007) used the same enzyme to transesterify *Jatropha* oil in $SC\text{-}CO_2$, and the highest conversions were in the range of 60% to 70%. Biodiesel was synthesized from mustard, linseed (Varma and Madras, 2007), and sesame oils (Varma et al., 2010) using Novozym®435 with different acyl acceptors in $SC\text{-}CO_2$. Mustard oil conversions were about 70% and 65% using methanol and ethanol, respectively. Slightly lower conversions were obtained from linseed oils with the highest yields being 45% and 35% in methanol and ethanol, respectively. A similar yield to that of linseed oils was obtained from animal fat at 50°C, 200 bar, a 4:1 molar ratio, and with a 30% loading of the lipase enzyme after 24 h (Taher et al., 2011). With sesame oil, the use of ethanol resulted in a yield of 55%, with methanol at 45%. A comprehensive study was conducted on the transesterification of microalgae lipids extracted by SC-CO$_2$ and converted by Novozyme®435. The effect of enzyme loading (15–50 wt%), temperature (35°C–55°C), and methanol-to-lipid (M:L) molar ratio (3–15:1) was investigated, and an optimum conversion yield of 82% was obtained at 47°C, 200 bar, 35% enzyme loading, and with a 9:1 M:L molar ratio after 4 h (Taher et al., 2014b).

6.6 HIGH-PRESSURE REACTION SYSTEMS

The two common modes of reactor operations are batch and continuous modes. Batch reactors are commonly used for kinetics studies, which are important for process optimization and scale-up. System design in this mode is simple, controllable, and cheap. Typically, reaction substrates are first added to the reactor, followed by the immobilized lipase. CO$_2$ is then introduced to the reactor and pressurized to the desired level. Figure 6.3 shows a schematic diagram of a setup used for enzymatic transesterification of lamb fats and microalgae lipids to biodiesel using Novozym®435 under SC-CO$_2$ (Taher, 2009, 2014).

Taher et al. (2011, 2014b) studied the effect of reaction temperature (35°C–60°C), methanol-to-lipids molar ratios (3:1–15:1), and enzyme loading (10%–50%) on biodiesel production yields from animal fats and microalgae lipids. Process conditions were optimized via response surface methodology. Transesterification yield of only 40% was obtained when animal fats were used, at 50°C, 200 bar, a 4:1 molar ratio, and a 30% loading of the enzyme (Taher et al., 2011), whereas a conversion above

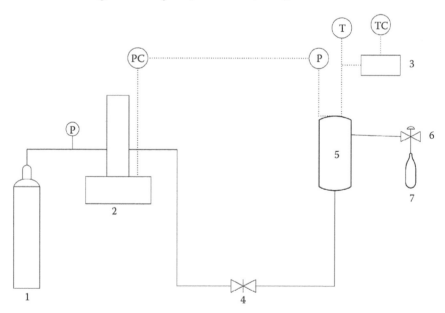

FIGURE 6.3 Schematic diagram of apparatus used for enzymatic reaction of lipids under SC-CO$_2$: 1, CO$_2$ cylinder with dip tube; 2, pump controller; 3, temperature controller; 4, prereaction valve; 5, reaction cell; 6, metering valve; 7, collection vial.

80% was achieved when microalgae lipids were used under the same operation conditions and a 9:1 molar ratio (Taher et al., 2014b). The lower molar ratio found when lamb fats were used was attributed to their solid nature and the low solubility of methanol in a highly saturated fat–SC-CO$_2$ mixture, compared to when unsaturated microalgae lipids were used. In addition, the diffusion of the microalgae lipids was higher than that of the lamb fats. This is mainly due to the negative effect of C18 unsaturated fatty acids that account for more than 75% and 50% of total content in microalgae lipids and lamb fat, respectively (Taher et al., 2014b).

At the end of the reaction, the reactor is depressurized by releasing SC-CO$_2$, which contains the biodiesel product. Depressurizing has a negative effect on enzyme activity and so limits the time the enzyme can be used for. To overcome this problem, continuous reactors are usually used. In semicontinuous flow reactors, an SC-CO$_2$ stream saturated with substrate flows continuously through a bed of immobilized lipase. Continuous mode processes are commonly used due to their ease of operation and continuous biodiesel recovery. In addition, the enzymes in such systems are not subjected to depressurization, which enhances its solubility and reuse (Laudani et al., 2007). Furthermore, bed regeneration can be achieved by washing with a suitable solvent, such as *tert*-butanol, that can dissolve the glycerol that accumulated in the bed. Dalla Rosa et al. (2009) investigated the continuous production of biodiesel from soybean oil in SC-CO$_2$. Ciftci and Temelli (2011, 2013) also investigated the conversion from corn oil in a similar system. Lubary et al. (2009) and Rodrigues et al. (2011) used a similar reaction process for biodiesel production from milk fat and sunflower oil, respectively.

6.7 PROCESSES INTEGRATION

Although high biodiesel yields are reported with continuous biodiesel production in SC-CO$_2$, the feasibility of such processes may not be obvious due to the high cost of pumping, despite reducing inhibition and easy product separation. Due to the advantages of SC-CO$_2$ over conventional organic solvents, the application of the high cost SC-CO$_2$ process may be justified in oil extraction. However, its justification for biodiesel production is not evident, despite certain positive effects. Nevertheless, it would be feasible to have a combined continuous process of extracting oils using SC-CO$_2$ and using the extracted oil for biodiesel production, using immobilized lipase in SC-CO$_2$, in one integrated system. In a continuous integrated process, the oils extracted are already dissolved in SC-CO$_2$ and can be fed directly into the enzymatic bioreactor to produce biodiesel without the need for further expensive pumping. In this way, the advantages of performing the reaction in a SC-CO$_2$ media are gained, while avoiding high pumping costs (Taher et al., 2014b). Figure 6.4 shows a schematic diagram of an integrated system used for continuous biodiesel production.

A lyophilized biomass was charged into the extraction cell, and the lipids extracted were mixed with a specific amount of methanol before entering the reaction cell, packed with Novozym®435. Both systems, extraction and reaction, were operated at 50°C and 200 bar. The proposed integrated system was tested at different methanol molar ratios, and Novozyme®435 operational stability and reuse were considered. After reaching a time when lipid extraction started to be limited by diffusion, the

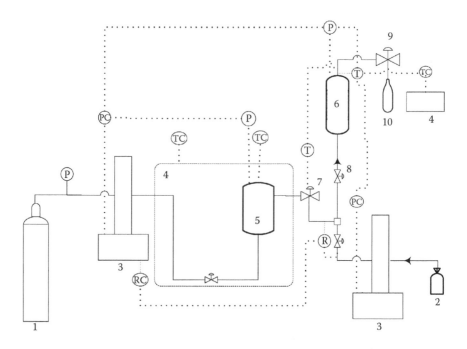

FIGURE 6.4 Schematic diagram of an integrated lipid extraction–reaction system used for continuous production: 1, CO_2 cylinder with dip tube; 2, methanol–*tert*-butanol mixture; 3, pump controller; 4, temperature controller; 5, extraction cell; 6, reaction cell; 7, enriched stream metering valve; 8, reaction mixture valve; 9, product emerging valve; and 10, product collection vial. (From Taher, H., S. Al-Zuhair, A. Al-Marzouqi, Y. Haik, and M. Farid, 2014, "Enzymatic Biodiesel Production of Microalgae Lipids under Supercritical Carbon Dioxide: Process Optimization and Integration," *Biochemical Engineering Journal* 90:103–113. With permission.)

biomass was replaced by a fresh sample, while the enzyme bed was unchanged. This process was repeated six times without a significant drop in enzyme activity. The bed stability, defined as a relative production rate in the sixth cycle to the maximum rate at each ratio, was found to be at 80% of its original activity. With a methanol:lipids molar ratio of 10:1, the maximum yield was obtained. At molar ratios above that, the yield dropped due to methanol inhibition. Due to the higher production yield, a slightly lower stability at the 10:1 ratio was found. This was due to the accumulation of excess glycerol. The accumulated glycerol formed a layer around the enzyme and inhibited the substrate from reaching the enzyme.

At the optimum molar ratio of 10:1, the enzyme bed was further utilized with fresh biomass in new cycles after washing the enzyme with *tert*-butanol. It was found that the *tert*-butanol dissolved the adsorbed glycerol and regenerated enzyme activity. The bed was successfully reused for six more cycles. In the second batch, after six cycles, the immobilized enzyme maintained up to 95% of its initial activity. The activity of the regenerated immobilized lipase at the end of each sixth cycle in the fourth batch after washing with *tert*-butanol was 93%.

REFERENCES

Abdullah, A. Z., B. Salamatinia, H. Mootabadi, and S. Bhatia. 2009. "Current Status and Policies on Biodiesel Industry in Malaysia as the World's Leading Producer of Palm Oil." *Energy Policy* 37 (12):5440–5448.

Adamczak, M., and S. H. Krishna. 2004. "Strategies for Improving Enzymes for Efficient Biocatalysis." *Food Technology and Biotechnology* 42 (4):251–264.

Adamczak, M., U. T. Bornscheuer, and W. Bednarski. 2009. "The Application of Biotechnological Methods for the Synthesis of Biodiesel." *European Journal of Lipid Science and Technology* 111:800–813.

Akoh, C. C., S.-W. Chang, G.-C. Lee, and J.-F. Shaw. 2007. "Enzymatic Approach to Biodiesel Production." *Journal of Agricultural and Food Chemistry* 55 (22):8995–9005.

Al-Zuhair, S. 2007. "Production of Biodiesel: Possibilities and Challenges." *Biofuels, Bioproducts and Biorefining* 1 (1):57–66.

Al-Zuhair, S., F. W. Ling, and L. S. Jun. 2007. "Proposed Kinetic Mechanism of the Production of Biodiesel from Palm Oil Using Lipase." *Process Biochemistry* 42 (6):951–960.

Antunes, W. M., C. de Oliveira Veloso, and C. A. Henriques. 2008. "Transesterification of Soybean Oil with Methanol Catalyzed by Basic Solids." *Catalysis Today* 133–135: 548–554.

Armenta, R. E., M. Vinatoru, A. M. Burja, J. A. Kralovec, and C. J. Barrow. 2007. "Transesterification of Fish Oil to Produce Fatty Acid Ethyl Esters Using Ultrasonic Energy." *Journal of the American Oil Chemists' Society* 84 (11):1045–1052.

Atadashi, I. M., M. K. Aroua, A. R. Abdul Aziz, and N. M. N. Sulaiman. 2012. "The Effects of Water on Biodiesel Production and Refining Technologies: A Review." *Renewable and Sustainable Energy Reviews* 16 (5):3456–3470.

Azam, M. M., A. Waris, and N. M. Nahar. 2005. "Prospects and Potential of Fatty Acid Methyl Esters of Some Non-Traditional Seed Oils for Use as Biodiesel in India." *Biomass and Bioenergy* 29 (4):293–302.

Bajpai, D., and V. K. Tyagi. 2006. "Biodiesel: Source, Production, Composition, Properties and Its Benifits." *Journal of Oleo Science* 55 (10):487–502.

Balat, M. 2011. "Potential Alternatives to Edible Oils for Biodiesel Production—A Review of Current Work." *Energy Conversion and Management* 52 (2):1479–1492.

Barbosa, D. da Costa, T. M. Serra, S. M. P. Meneghetti, and M. R. Meneghetti. 2010. "Biodiesel Production by Ethanolysis of Mixed Castor and Soybean Oils." *Fuel* 89 (12):3791–3794.

Basha, S. A., K. Raja Gopal, and S. Jebaraj. 2009. "A Review on Biodiesel Production, Combustion, Emissions and Performance." *Renewable Sustainable Energy Reviews* 13 (6–7):1628–1634.

Becker, W. 2004. "Microalgae in Human and Animal Nutrition." In *Handbook of Microalgal Culture: Biotechnology and Applied Phycology*, edited by A. Richmond. Blackwell Science.

Berchmans, H. J., and S. Hirata. 2008. "Biodiesel Production from Crude *Jatropha Curcas* L. Seed Oil with a High Content of Free Fatty Acids." *Bioresource Technology* 99 (6):1716–1721.

Bernardes, O. L., J. V. Bevilaqua, M. C. M. R. Leal, D. M. G. Freire, and M. A. P. Langone. 2007. "Biodiesel Fuel Production by the Transesterification Reaction of Soybean Oil Using Immobilized Lipase." *Applied Biochemistry and Biotechnology* 137–140 (1–12):105–114.

Caballero, V., F. M. Bautista, J. M. Campelo, D. Luna, J. M. Marinas, A. A. Romero, J. M. Hidalgo, R. Luque, A. Macario, and G. Giordano. 2009. "Sustainable Preparation of a Novel Glycerol-Free Biofuel by Using Pig Pancreatic Lipase: Partial 1,3-Regiospecific Alcoholysis of Sunflower Oil." *Process Biochemistry* 44 (3):334–342.

Canakci, M., and J. Van Gerpen. 2001. "Biodiesel Production from Oils and Fat with High Free Fatty Acids." *American Society of Agricultural Engineers* 44 (6):1429–1436.

Carelli, A. A., M. I. V. Brevedan, and G. H. Crapiste. 1997. "Quantitative Determination of Phospholipids in Sunflower Oil." *Journal of the American Oil Chemists' Society* 74 (5):511–514.

Chang, H.-M., H.-F. Liao, C.-C. Lee, and C.-J. Shieh. 2005. "Optimized Synthesis of Lipase-Catalyzed Biodiesel by Novozym 435." *Journal of Chemical Technology & Biotechnology* 80 (3):307–312.

Chen, G., M. Ying, and W. Li. 2006. "Enzymatic Conversion of Waste Cooking Oils into Alternative Fuel-Biodiesel." *Applied Biochemistry and Biotechnology* 132 (1–3): 911–921.

Chen, H.-C., H.-Y. Ju, T.-T. Wu, Y.-C. Liu, C.-C. Lee, C. Chang, Y.-L. Chung, and C.-J. Shieh. 2011. "Continuous Production of Lipase-Catalyzed Biodiesel in a Packed-Bed Reactor: Optimization and Enzyme Reuse Study." *Journal of Biomedicine and Biotechnology* 2011 (Article ID 950725).

Chen, L., T. Liu, W. Zhang, X. Chen, and J. Wang. 2012. "Biodiesel Production from Algae Oil High in Free Fatty Acids by Two-Step Catalytic Conversion." *Bioresource Technology* 111:208–214.

Chisti, Y. 2007. "Biodiesel from Microalgae." *Biotechnology Advances* 25 (3):294–306.

Chung, K.-H., J. Kim, and K.-Y. Lee. 2009. "Biodiesel Production by Transesterification of Duck Tallow with Methanol on Alkali Catalysts." *Biomass and Bioenergy* 33 (1):155–158.

Ciftci, O. N., and F. Temelli. 2011. "Continuous Production of Fatty Acid Methyl Esters from Corn Oil in a Supercritical Carbon Dioxide Bioreactor." *Journal of Supercritical Fluids* 58 (1):79–87.

Ciftci, O. N., and F. Temelli. 2013. "Continuous Biocatalytic Conversion of the Oil of Corn Distiller's Dried Grains with Solubles to Fatty Acid Methyl Esters in Supercritical Carbon Dioxide." *Biomass and Bioenergy* 54 (0):140–146.

Dalla Rosa, C., M. B. Morandim, J. L. Ninow, D. Oliveira, H. Treichel, and J. Vladimir Oliveira. 2009. "Continuous Lipase-Catalyzed Production of Fatty Acid Ethyl Esters from Soybean Oil in Compressed Fluids." *Bioresource Technology* 100 (23):5818–5826.

Demirbas, A. 1998. "Fuel Properties and Calculation of Higher Heating Values of Vegetable Oils." *Fuel* 77 (9–10):1117–1120.

Demirbas, A. 2002. "Biodiesel from Vegetable Oils via Transesterification in Supercritical Methanol." *Energy Conversion and Management* 43 (17):2349–2356.

Demirbas, A. 2003. "Biodiesel Fuels from Vegetable Oils via Catalytic and Non-Catalytic Supercritical Alcohol Transesterifications and Other Methods: A Survey." *Energy Conversion and Management* 44 (13):2093–2109.

Demirbas, A. 2007a. "Importance of Biodiesel as Transportation Fuel." *Energy Policy* 35 (9):4661–4670

Demirbas, A. 2007b. "Progress and Recent Trends in Biofuels." *Progress in Energy and Combustion Science* 33 (1):1–18.

Demirbas, A. 2009a. "Biodiesel from Waste Cooking Oil via Base-Catalytic and Supercritical Methanol Transesterification." *Energy Conversion and Management* 50 (4):923–927.

Demirbas, A. 2009b. "Progress and Recent Trends in Biodiesel Fuels." *Energy Conversion and Management* 50:14–34.

Demirbas, A. 2010. "Use of Algae as Biofuel Sources." *Energy Conversion and Management* 51 (12):2738–2749.

Demirbas, A., and M. F. Demirbas. 2011. "Importance of Algae Oil as a Source of Biodiesel." *Energy Conversion and Management* 52:163–170.

Demirel, G., H. Ozpinar, B. Nazli, and O. Keser. 2006. "Fatty Acids of Lamb Meat from Two Breeds Fed Different Forage: Concentrate Ratio." *Meat Science* 72 (2):229–235.

Devanesan, M. G., T. Viruthagiri, and N. Sugumar. 2007. "Transesterification of Jatropha Oil Using Immobilized Pseudomonas Fluorescens." *African Journal of Biotechnology* 6 (21):2497–2501.

Dossin, T. F., M.-F. Reyniers, R. J. Berger, and G. B. Marin. 2006. "Simulation of Heterogeneously MgO-Catalyzed Transesterification for Fine-Chemical and Biodiesel Industrial Production." *Applied Catalysis B: Environmental* 67 (1–2):136–148.

Doukyu, N., and H. Ogino. 2010. "Organic Solvent-Tolerant Enzymes." *Biochemical Engineering Journal* 48 (3):270–282.

Du, W., Y. Xu, D. Liu, and J. Zeng. 2004. "Comparative Study on Lipase-Catalyzed Transformation of Soybean Oil for Biodiesel Production with Different Acyl Acceptors." *Journal of Molecular Catalysis B: Enzymatic* 30 (3–4):125–129.

Dubé, M. A., A. Y. Tremblay, and J. Liu. 2007. "Biodiesel Production Using a Membrane Reactor." *Bioresource Technology* 98 (3):639–647

Ebiura, T., T. Echizen, A. Ishikawa, K. Murai, and T. Baba. 2005. "Selective Transesterification of Triolein with Methanol to Methyl Oleate and Glycerol Using Alumina Loaded with Alkali Metal Salt as a Solid-Base Catalyst." *Applied Catalysis A: General* 283 (1–2):111–116.

Eltaweel, M. A., R. N. Z. R. Abd Rahman, A. B. Salleh, and M. Basri. 2005. "An Organic Solvent-Stable Lipase from *Bacillus* Sp. Strain 42." *Annals of Microbiology* 55 (3):187–192.

Encinar, J. M., J. F. Gonzalez, and A. Rodriguez-Reinares. 2007. "Ethanolysis of Used Frying Oil. Biodiesel Preparation and Characterization." *Fuel Processing Technology* 88 (5):513–522.

Fábregas, J., A. Maseda, A. Domínguez, and A. Otero. 2004. "The Cell Composition of *Nannochloropsis* Sp. Changes under Different Irradiances in Semicontinuous Culture." *World Journal of Microbiology and Biotechnology* 20 (1):31–35.

Fadhil, A. B., M. M. Dheyab, K. M. Ahmed, and M. H. Yahyaa. 2012. "Biodiesel Production from Spent Fish Frying Oil through Acid-Base Catalyzed Transesterification." *Pakistan Journal of Analytical & Environmental Chemistry* 13 (1):9–15.

Fan, X., and R. Burton. 2009. "Recent Development of Biodiesel Feedstocks and the Applications of Glycerol: A Review." *Open Fuels & Energy Science Journal* 1:100–109.

Fassinou, W. F., A. Sako, A. Fofana, K. B. Koua, and S. Toure. 2010. "Fatty Acids Composition as a Means to Estimate the High Heating Value (HHV) of Vegetable Oils and Biodiesel Fuels." *Energy* 35 (12):4949–4954.

Fellows, P. 2000. *Food Processing Technology: Principles and Practice.* Woodhead Publishing and CRC Press.

Fjerbaek, L., K. V. Christensen, and B. Norddahl. 2009. "A Review of the Current State of Biodiesel Production Using Enzymatic Transesterification." *Biotechnology and Bioengineering* 102 (5):1298–1315.

Freedman, B., E. H. Pryde, and T. L. Mounts. 1984. "Variables Affecting the Yields of Fatty Esters from Transesterified Vegetable Oils." *Journal of the American Oil Chemists Society* 61 (10):1638–1643.

Fukuda, H., A. Kondo, and H. Noda. 2001. "Biodiesel Fuel Production by Transesterification of Oils." *Journal of Bioscience and Bioengineering* 92 (5):405–416.

Georgogianni, K. G., M. G. Kontominas, P. J. Pomonis, D. Avlonitis, and V. Gergis. 2008. "Conventional and in Situ Transesterification of Sunflower Seed Oil for the Production of Biodiesel." *Fuel Processing Technology* 89 (5):503–509.

Ghadge, S. V., and H. Raheman. 2005. "Biodiesel Production from Mahua (*Madhuca Indica*) Oil Having High Free Fatty Acids." *Biomass and Bioenergy* 28 (6):601–605.

Giessauf, A., and T. Gamse. 2000. "A Simple Process for Increasing the Specific Activity of Porcine Pancreatic Lipase by Supercritical Carbon Dioxide Treatment." *Journal of Molecular Catalysis B: Enzymatic* 9 (1–3):57–64.

Graboski, M. S., and R. L. McCormick. 1998. "Combustion of Fat and Vegetable Oil Derived Fuels in Diesel Engines." *Progress in Energy and Combustion Science* 24 (2):125–164.

Graham, L. E., and L. W. Wilcox. 2000. *Algae*. Prentice Hall.

Granados, M. L., M. D. Zafra Poves, D. Martín Alonso, R. Mariscal, F. Cabello Galisteo, R. Moreno-Tost, J. Santamaría, and J. L. G. Fierro. 2007. "Biodiesel from Sunflower Oil by Using Activated Calcium Oxide." *Applied Catalysis B: Environmental* 73 (3–4):317–326.

Guan, G., K. Kusakabe, and S. Yamasaki. 2009. "Tri-Potassium Phosphate as a Solid Catalyst for Biodiesel Production from Waste Cooking Oil." *Fuel Processing Technology* 90 (4):520–524.

Ha, S. H., M. N. Lan, S. H. Lee, S. M. Hwang, and Y.-M. Koo. 2007. "Lipase-Catalyzed Biodiesel Production from Soybean Oil in Ionic Liquids." *Enzyme and Microbial Technology* 41 (4):480–483.

Haas, M. J., and T. A. Foglia. 2005. "Alternate Feedstocks and Technologies for Biodiesel Production." In *The Biodiesel Handbook*, edited by G. Knothe, J. Van Gerpen, and J. Krahl, 42–61. AOCS Press.

Harwood, J. L., and I. A. Guschina. 2009. "The Versatility of Algae and Their Lipid Metabolism." *Biochimie* 91 (6):679–684.

Helwani, Z., M. R. Othman, N. Aziz, W. J. N. Fernando, and J. Kim. 2009. "Technologies for Production of Biodiesel Focusing on Green Catalytic Techniques: A Review." *Fuel Processing Technology* 90 (12):1502–1514.

Hernández-Rodríguez, B., J. Córdova, E. Bárzana, and E. Favela-Torres. 2009. "Effects of Organic Solvents on Activity and Stability of Lipases Produced by Thermotolerant Fungi in Solid-State Fermentation." *Journal of Molecular Catalysis B: Enzymatic* 61 (3–4):136–142.

Hess, M. A., M. J. Haas, T. A. Foglia, and W. N. Marmer. 2005. "Effect of Antioxidant Addition on NO_x Emissions from Biodiesel." *Energy Fuels* 19 (4):1749–1754.

Hsu, A.-F., K. Jones, W. Marmer, and T. Foglia. 2001. "Production of Alkyl Esters from Tallow and Grease Using Lipase Immobilized in a Phyllosilicate Sol-Gel." *Journal of the American Oil Chemists' Society* 78 (6):585–588.

Huang, Z., X.-H. Shi, and W.-J. Jiang. 2012. "Theoretical Models for Supercritical Fluid Extraction." *Journal of Chromatography A* 1250:2–26.

Iso, M., B. Chen, M. Eguchi, T. Kudo, and S. Shrestha. 2001. "Production of Biodiesel Fuel from Triglycerides and Alcohol Using Immobilized Lipase." *Journal of Molecular Catalysis B: Enzymatic* 16 (1):53–58.

Jacobson, K., R. Gopinath, L. C. Meher, and A. K. Dalai. 2008. "Solid Acid Catalyzed Biodiesel Production from Waste Cooking Oil." *Applied Catalysis B: Environmental* 85 (1–2):86–91.

Jegannathan, K. R., L. Jun-Yee, E.-S. Chan, and P. Ravindra. 2009. "Design an Immobilized Lipase Enzyme for Biodiesel Production." *Journal of Renewable and Sustainable Energy* 1 (6):063101.

Jegannathan, K. R., C. Eng-Seng, and P. Ravindra. 2011. "Economic Assessment of Biodiesel Production: Comparison of Alkali and Biocatalyst Processes." *Renewable and Sustainable Energy Reviews* 15 (1):745–751.

Ji, J., J. Wang, Y. Li, Y. Yu, and Z. Xu. 2006. "Preparation of Biodiesel with the Help of Ultrasonic and Hydrodynamic Cavitation." *Ultrasonics* 44 (Suppl. 1):e411–e414.

Kaieda, M., T. Samukawa, A. Kondo, and H. Fukuda. 2001. "Effect of Methanol and Water Contents on Production of Biodiesel Fuel from Plant Oil Catalyzed by Various Lipases in a Solvent-Free System." *Journal of Bioscience and Bioengineering* 91 (1):12–15.

Kalam, M. A., and H. H. Masjuki. 2002. "Biodiesel from Palmoil—An Analysis of Its Properties and Potential." *Biomass Bioenergy* 23 (6):471–479.

Kamal-Eldin, A., and R. Andersson. 1997. "A Multivariate Study of the Correlation between Tocopherol Content and Fatty Acid Composition in Vegetable Oils." *Journal of the American Oil Chemists' Society* 74 (4):375–380.

Karmee, S. K., and A. Chadha. 2005. "Preparation of Biodiesel from Crude Oil of *Pongamia Pinnata*." *Bioresource Technology* 96 (13):1425–1429.

Klibanov, A. M. 1997. "Why Are Enzymes Less Active in Organic Solvents Than in Water?" *Trends in Biotechnology* 15 (3):97–101.

Klibanov, A. M. 2001. "Improving Enzymes by Using Them in Organic Solvents." *Nature* 209:241–246.

Knauer, J., and P. C. Southgate. 1999. "A Review of the Nutritional Requirements of Bivalves and the Development of Alternative and Artificial Diets for Bivalve Aquaculture." *Reviews in Fisheries Science* 7 (3):241–280.

Knothe, G. 2005. "Dependence of Biodiesel Fuel Properties on the Structure of Fatty Acid Alkyl Esters." *Fuel Processing Technology* 86 (10):1059–1070.

Knothe, G. 2009. "Improving Biodiesel Fuel Properties by Modifying Fatty Ester Composition." *Energy & Environmental Science* 2 (7):759–766.

Koh, M. Y., and T. I. M. Ghazi. 2011. "A Review of Biodiesel Production from *Jatropha Curcas* L. Oil." *Renewable and Sustainable Energy Reviews* 15 (5):2240–2251.

Köse, Ö., M. Tüter, and H. A. Aksoy. 2002. "Immobilized Candida Antarctica Lipase-Catalyzed Alcoholysis of Cotton Seed Oil in a Solvent-Free Medium." *Bioresource Technology* 83 (2):125–129

Kouzu, M., T. Kasuno, M. Tajika, Y. Sugimoto, S. Yamanaka, and J. Hidaka. 2008. "Calcium Oxide as a Solid Base Catalyst for Transesterification of Soybean Oil and Its Application to Biodiesel Production." *Fuel* 87 (12):2798–2806.

Kucek, K. T., M. Aparecida F. Cesar-Oliveira, H. M. Wilhelm, and L. P. Ramos. 2007. "Ethanolysis of Refined Soybean Oil Assisted by Sodium and Potassium Hydroxides." *Journal of the American Oil Chemists' Society* 84 (4):385–392.

Kumar, R., G. Madras, and J. Modak. 2004. "Enzymatic Synthesis of Ethyl Palmitate in Supercritical Carbon." *Industrial & Engineering Chemistry Research* 43 (7):1568–1573.

Kumari, V., S. Shah, and M. N. Gupta. 2007. "Preparation of Biodiesel by Lipase-Catalyzed Transesterification of High Free Fatty Acid Containing Oil from Madhuca Indica." *Energy & Fuels* 21 (1):368–372.

Lai, C.-C., S. Zullaikah, S. R. Vali, and Y.-H. Ju. 2005. "Lipase-Catalyzed Production of Biodiesel from Rice Bran Oil." *Journal of Chemical Technology & Biotechnology* 80 (3):331–337.

Lam, M. K., K. T. Lee, and A. R. Mohamed. 2009. "Sulfated Tin Oxide as Solid Superacid Catalyst for Transesterification of Waste Cooking Oil: An Optimization Study." *Applied Catalysis B: Environmental* 93 (1–2):134–139.

Lam, M. K., K. T. Lee, and A. R. Mohamed. 2010. "Homogeneous, Heterogeneous and Enzymatic Catalysis for Transesterification of High Free Fatty Acid Oil (Waste Cooking Oil) to Biodiesel: A Review." *Biotechnology Advances* 28 (4):500–518.

Laudani, C. G., M. Habulin, Ž. Knez, G. D. Porta, and E. Reverchon. 2007. "Lipase-Catalyzed Long Chain Fatty Ester Synthesis in Dense Carbon Dioxide: Kinetics and Thermodynamics." *Journal of Supercritical Fluids* 41 (1):92–101

Lee, K.-T., T. A. Foglia, and K.-S. Chang. 2002. "Production of Alkyl Ester as Biodiesel from Fractionated Lard and Restaurant Grease." *Journal of the American Oil Chemists' Society* 79 (2):191–195.

Lee, M., D. Lee, Ja. Cho, J. Lee, S. Kim, S. W. Kim, and C. Park. 2013. "Optimization of Enzymatic Biodiesel Synthesis Using RSM in High Pressure Carbon Dioxide and Its Scale Up." *Bioprocess and Biosystems Engineering* 36 (6):775–780.

Leung, D. Y. C., and Y. Guo. 2006. "Transesterification of Neat and Used Frying Oil: Optimization for Biodiesel Production." *Fuel Processing Technology* 87 (10):8 83–890.

Li, L., W. Du, D. Liu, L. Wang, and Z. Li. 2006. "Lipase-Catalyzed Transesterification of Rapeseed Oils for Biodiesel Production with a Novel Organic Solvent as the Reaction Medium." *Journal of Molecular Catalysis B: Enzymatic* 43 (1–4):58–62

Li, X., H. Xu, and Q. Wu. 2007. "Large-Scale Biodiesel Production from Microalga Chlorella Protothecoides through Heterotrophic Cultivation in Bioreactors." *Biotechnology and Bioengineering* 98 (4):764–771.

Lin, L., D. Ying, S. Chaitep, and S. Vittayapadung. 2009. "Biodiesel Production from Crude Rice Bran Oil and Properties as Fuel." *Applied Energy* 86:681–688.

Liu, Y., E. Lotero, and J. G. Goodwin Jr. 2006. "Effect of Water on Sulfuric Acid Catalyzed Esterification." *Journal of Molecular Catalysis A: Chemical* 245 (1–2):132–140.

Liu, X., H. He, Y. Wang, S. Zhu, and X. Piao. 2008. "Transesterification of Soybean Oil to Biodiesel Using Cao as a Solid Base Catalyst." *Fuel* 87 (2):216–221.

Lu, J., K. Nie, F. Xie, F. Wang, and T. Tan. 2007. "Enzymatic Synthesis of Fatty Acid Methyl Esters from Lard with Immobilized Candida Sp. 99-125." *Process Biochemistry* 42 (9):1367–1370.

Lubary, M., P. J. Jansens, J. H. ter Horst, and G. W. Hofland. 2009. "Integrated Synthesis and Extraction of Short-Chain Fatty Acid Esters by Supercritical Carbon Dioxide." *AIChE Journal* 56 (4):1080–1089.

Ma, F., and M. A Hanna. 1999. "Biodiesel Production: A Review." *Bioresource Technology* 70 (1):1–15.

Ma, F., L. D. Clements, and M. A. Hanna. 1998. "Biodiesel Fuel from Animal Fat. Ancillary Studies on Transesterification of Beef Tallow." *Industrial and Engineering Chemistry Research* 37 (9):3768–3771.

Madras, G., C. Kolluru, and R. Kumar. 2004. "Synthesis of Biodiesel in Supercritical Fluids." *Fuel* 83 (14–15):2029–2033.

Marchetti, J. M., V. U. Miguel, and A. F. Errazu. 2007. "Possible Methods for Biodiesel Production." *Renewable and Sustainable Energy Reviews* 11 (6):1300–1311.

Marjanovic, A. V., O. S. Stamenkovic, Z. B. Todorovic, M. L. Lazic, and V. B. Veljkovic. 2010. "Kinetics of the Base-Catalyzed Sunflower Oil Ethanolysis." *Fuel* 89 (3):665–671.

Marmesat, S., E. Rodrigues, J. Velasco, and C. Dobarganes. 2007. "Quality of Used Frying Fats and Oils: Comparison of Rapid Tests Based on Chemical and Physical Oil Properties." *International Journal of Food Science & Technology* 42 (5):601–608.

Meher, L. C., V. S. S. Dharmagadda, and S. N. Naik. 2006. "Optimization of Alkali-Catalyzed Transesterification of Pongamia Pinnata Oil for Production of Biodiesel." *Bioresource Technology* 97 (12):1392–1397.

Meher, L. C., M. G. Kulkarni, A. K. Dalai, and S. N. Naik. 2006. "Transesterification of Karanja (*Pongamia Pinnata*) Oil by Solid Basic Catalysts." *European Journal of Lipid Science and Technology* 108 (5):389–397.

Mendow, G., N. S. Veizaga, and C. A. Querini. 2011. "Ethyl Ester Production by Homogeneous Alkaline Transesterification: Influence of the Catalyst." *Bioresource Technology* 102 (11):6385–6391.

Meneghetti, S. M. P., M. R. Meneghetti, C. R. Wolf, E. C. Silva, G. E. S. Lima, L. de Lira Silva, T. M. Serra, F. Cauduro, and L. G. de Oliveira. 2006. "Biodiesel from Castor Oil: A Comparison of Ethanolysis Versus Methanolysis." *Energy & Fuels* 20 (5):2262–2265.

Meng, X., G. Chen, and Y. Wang. 2008. "Biodiesel Production from Waste Cooking Oil Via Alkali Catalyst and Its Engine Test." *Fuel Processing Technology* 89 (9):851–857.

Meng, X., J. Yang, X. Xu, L. Zhang, Q. Nie, and M. Xian. 2009. "Biodiesel Production from Oleaginous Microorganisms." *Renewable Energy* 34 (1):1–5.

Miao, X., and Q. Wu. 2004. "High Yield Bio-Oil Production from Fast Pyrolysis by Metabolic Controlling of Chlorella Protothecoides." *Journal of Biotechnology* 110 (1):85–93.

Miao, X., and Q. Wu. 2006. "Biodiesel Production from Heterotrophic Microalgal Oil." *Bioresource Technology* 97 (6):841–846.

Miao, X., Q. Wu, and C. Yang. 2004. "Fast Pyrolysis of Microalgae to Produce Renewable Fuels." *Journal of Analytical and Applied Pyrolysis* 71 (2):855–863.

Mittelbach, M. 1990. "Lipase Catalyzed Alcoholysis of Sunflower Oil." *Journal of the American Oil Chemists' Society* 67 (3):168–170.

Modi, M. K., J. R. C. Reddy, B. V. S. K. Rao, and R. B. N. Prasad. 2006. "Lipase-Mediated Transformation of Vegetable Oils into Biodiesel Using Propan-2-Ol as Acyl Acceptor." *Biotechnology Letters* 28 (9):637–640.

Modi, M. K., J. R. C. Reddy, B. V. S. K. Rao, and R. B. N. Prasad. 2007. "Lipase-Mediated Conversion of Vegetable Oils into Biodiesel Using Ethyl Acetate as Acyl Acceptor." *Bioresource Technology* 98 (6):1260–1264

Moser, B. R., and S. F. Vaughn. 2010. "Coriander Seed Oil Methyl Esters as Biodiesel Fuel: Unique Fatty Acid Composition and Excellent Oxidative Stability." *Biomass and Bioenergy* 34 (4):550–558.

Naik, S. N., V. V. Goud, P. K. Rout, and A. K. Dalai. 2010. "Production of First and Second Generation Biofuels: A Comprehensive Review." *Renewable and Sustainable Energy Reviews* 14 (2):578–597.

Natrah, F., F. Yusoff, M. Shariff, F. Abas, and N. Mariana. 2007. "Screening of Malaysian Indigenous Microalgae for Antioxidant Properties and Nutritional Value." *Journal of Applied Phycology* 19 (6):711–718.

Navarro-Diaz, H. J., S. L. Gonzalez, B. Irigaray, I. Vieitez, I. Jachmanián, H. Hense, and J. V. Oliveira. 2014. "Macauba Oil as an Alternative Feedstock for Biodiesel: Characterization and Ester Conversion by the Supercritical Method." Paper presented at III Iberoamerican Conference on Supercritical Fluids—PROSCIBA 2013, Cartagena, Colombia, April 1–5.

Nelson, L. A., T. A. Foglia, and W. N. Marmer. 1996. "Lipase-Catalyzed Production of Biodiesel." *Journal of Americal Oil Chemists' Society* 73 (9):1191–1195.

Noureddini, H., X. Gao, and R. S. Philkana. 2005. "Immobilized Pseudomonas Cepacia Lipase for Biodiesel Fuel Production from Soybean Oil." *Bioresource Technology* 96 (7):769–777.

Novak, Z., M. Habulin, V. Krmelj, and Z. Knez. 2003. "Silica Aerogels as Supports for Lipase Catalyzed Esterifications at Sub- and Supercritical Conditions." *Journal of Supercritical Fluids* 27 (2):169–178.

Oliveira, D., and J. Vladimir Oliveira. 2001. "Enzymatic Alcoholysis of Palm Kernel Oil in *n*-Hexane and SCCO$_2$." *Journal of Supercritical Fluids* 19:141–148.

Orçaire, O., P. Buisson, and A. C. Pierre. 2006. "Application of Silica Aerogel Encapsulated Lipases in the Synthesis of Biodiesel by Transesterification Reactions." *Journal of Molecular Catalysis B: Enzymatic* 42 (3–4):106-113.

Park, Y.-M., D.-W. Lee, D.-K. Kim, J.-S. Lee, and K.-Y. Lee. 2008. "The Heterogeneous Catalyst System for the Continuous Conversion of Free Fatty Acids in Used Vegetable Oils for the Production of Biodiesel." *Catalysis Today* 131 (1–4):238–243.

Patil, V., K.-Q. Tran, and H. R. Giselrød. 2008. "Towards Sustainable Production of Biofuels from Microalgae." *International Journal of Molecular Sciences* 9:1188–1195.

Patil, P., S. Deng, J. I. Rhodes, and P. J. Lammers. 2010. "Conversion of Waste Cooking Oil to Biodiesel Using Ferric Sulfate and Supercritical Methanol Processes." *Fuel* 89 (2):360–364.

Peng, B.-X., Q. Shu, J.-F. Wang, G.-R. Wang, D.-Z. Wang, and M.-H. Han. 2008. "Biodiesel Production from Waste Oil Feedstocks by Solid Acid Catalysis." *Process Safety and Environmental Protection* 86 (6):441–447.

Pereyra-Irujo, G. A., N. G. Izquierdo, M. Covi, S. M. Nolascoc, F. Quiroza, and L. A. N. Aguirrezábal. 2009. "Variability in Sunflower Oil Quality for Biodiesel Production: A Simulation Study." *Biomass Bioenergy* 33 (3):459–468.

Peterson, G., and W. Scarrah. 1984. "Rapeseed Oil Transesterification by Heterogeneous Catalysis." *Journal of the American Chemical Society* 61 (10):1593–1597.

Phan, A. N., and T. M. Phan. 2008. "Biodiesel Production from Waste Cooking Oils." *Fuel* 87 (17–18):3490–3496.

Pokoo-Aikins, G., A. Nadim, M. El-Halwagi, and V. Mahalec. 2009. "Design and Analysis of Biodiesel Production from Algae Grown through Carbon Sequestration." *Clean Technologies and Environmental Policy* 12 (3):239–254.

Radzi, S., M. Basri, A. Salleh, A. Ariff, R. Mohammad, M. B. Abdul Rahman, and R. N. Z. A. Rahman. 2005. "High Performance Enzymatic Synthesis of Oleyl Oleate Using Immobilised Lipase from *Candida Antartica*." *Electronic Journal of Biotechnology* 8 (3).

Rajan, K., and K. R. Senthilkumar. 2009. "Effect of Exhaust Gas Recirculation (EGR) on the Performance and Emission Characteristics of Diesel Engine with Sunflower Oil Methyl Ester." *Jordan Journal of Mechanical and Industrial Engineering* 3 (4):306–311.

Ramos, M. J., C. M. Fernández, A. Casas, L. Rodríguez, and Á. Pérez. 2009. "Influence of Fatty Acid Composition of Raw Materials on Biodiesel Properties." *Bioresource Technology* 100 (1):261–268.

Ranganathan, S. V., S. L. Narasimhan, and K. Muthukumar. 2008. "An Overview of Enzymatic Production of Biodiesel." *Bioresource Technology* 99 (10):3975–3981.

Rashid, U., F. Anwar, B. R. Moser, and S. Ashraf. 2008. "Production of Sunflower Oil Methyl Esters by Optimized Alkali-Catalyzed Methanolysis." *Biomass and Bioenergy* 32 (12):1202–1205.

Rathore, V., and G. Madras. 2007. "Synthesis of Biodiesel from Edible and Non-Edible Oils in Supercritical Alcohols and Enzymatic Synthesis in Supercritical Carbon Dioxide." *Fuel* 86 (17–18):2650–2659.

Refaat, A. A., N. K. Attia, H. A. Sibak, S. T. Sheltawy, and G. I. ElDiwani. 2008. "Production Optimization and Quality Assessment of Biodiesel from Waste Vegetable Oil." *International Journal of Environmental Science & Technology* 5 (1):75–82.

Renaud, S. M., L.-V. Thinh, G. Lambrinidis, and D. L. Parry. 2002. "Effect of Temperature on Growth, Chemical Composition and Fatty Acid Composition of Tropical Australian Microalgae Grown in Batch Cultures." *Aquaculture* 211 (1–4):195–214.

Repka, S., M. van der Vlies, and J. Vijverberg. 1998. "Food Quality of Detritus Derived from the Filamentous Cyanobacterium *Oscillatoria Limnetica* for *Daphnia Galeata*." *Journal of Plankton Research* 20 (11):2199–2205.

Robles-Medina, A., P. A. González-Moreno, L. Esteban-Cerdán, and E. Molina-Grima. 2009. "Biocatalysis: Towards Ever Greener Biodiesel Production." *Biotechnology Advances* 27 (4):398–408.

Rodrigues, A. R., A. Paiva, M. Gomes da Silva, P. Simões, and S. Barreiros. 2011. "Continuous Enzymatic Production of Biodiesel from Virgin and Waste Sunflower Oil in Supercritical Carbon Dioxide." *Journal of Supercritical Fluids* 56 (3):259–264.

Romero, M. D., L. Calvo, C. Alba, M. Habulin, M. Primozic, and Z. Knez. 2005. "Enzymatic Synthesis of Isoamyl Acetate with Immobilized *Candida Antarctica* Lipase in Supercritical Carbon Dioxide." *Journal of Supercritical Fluids* 33 (1):77–84.

Royon, D., M. Daz, G. Ellenrieder, and S. Locatelli. 2007. "Enzymatic Production of Biodiesel from Cotton Seed Oil Using t-Butanol as a Solvent." *Bioresource Technology* 98 (3):648–653

Ruzich, N. I., and A. S. Bassi. 2010. "Investigation of Enzymatic Biodiesel Production Using Ionic Liquid as a Co-Solvent." *Canadian Journal of Chemical Engineering* 88 (2):277–282.

Saka, S., and D. Kusdiana. 2001. "Biodiesel Fuel from Rapeseed Oil as Prepared in Supercritical Methanol." *Fuel* 80 (2):225–231.

Salis, A., M. Pinna, M. Monduzzi, and V. Solinas. 2005. "Biodiesel Production from Triolein and Short Chain Alcohols through Biocatalysis." *Journal of Biotechnology* 119 (3):291–299.

Salis, A., M. Pinna, M. Monduzzi, and V. Solinas. 2008. "Comparison among Immobilised Lipases on Macroporous Polypropylene toward Biodiesel Synthesis." *Journal of Molecular Catalysis B: Enzymatic* 54 (1–2):19–26.

Samniang, A., C. Tipachan, and S. Kajorncheappun-Ngam. 2014. "Comparison of Biodiesel Production from Crude Jatropha Oil and Krating Oil by Supercritical Methanol Transesterification." *Renewable Energy* 68:351–355.

Samukawa, T., M. Kaieda, T. Matsumoto, K. Ban, A. Kondo, Y. Shimada, H. Noda, and H. Fukuda. 2000. "Pretreatment of Immobilized *Candida Antarctica* Lipase for Biodiesel Fuel Production from Plant Oil." *Journal of Bioscience and Bioengineering* 90 (2):180–183.

Santambrogio, C., F. Sasso, A. Natalello, S. Brocca, R. Grandori, S. M. Doglia, and M. Lotti. 2013. "Effects of Methanol on a Methanol-Tolerant Bacterial Lipase." *Applied Microbiology and Biotechnology* 97 (19):8609–8618.

Sathya, T., and A. Manivannan. 2013. "Biodiesel Production from Neem Oil Using Two Step Transesterification." *International Journal of Engineering Research and Applications* 3 (3):488–492.

Satpati, G. G., and R. Pal. 2014. "Rapid Detection of Neutral Lipid in Green Microalgae by Flow Cytometry in Combination with Nile Red Staining—An Improved Technique." *Annals of Microbiology*:1–13.

Schumacher, L. G., W. Marshall, J. Krahl, and W. B. Wet. 2001. "Biodiesel Emissions Data from Series 60 DDC Engines." *Transactions of the ASAE* 44 (6):1465–1468.

Shah, S., and M. N. Gupta. 2007. "Lipase Catalyzed Preparation of Biodiesel from Jatropha Oil in a Solvent Free System." *Process Biochemistry* 42 (3):409–414.

Shah, S., S. Sharma, and M. N. Gupta. 2004. "Biodiesel Preparation by Lipase-Catalyzed Transesterification of Jatropha Oil." *Energy & Fuels* 18 (1):154–159.

Sharma, Y. C., and B. Singh. 2010. "A Hybrid Feedstock for a Very Efficient Preparation of Biodiesel." *Fuel Processing Technology* 91 (10):1267–1273.

Sharma, Y. C., B. Singh, and S. N. Upadhyay. 2008. "Advancements in Development and Characterization of Biodiesel: A Review." *Fuel* 87 (12):2355–2373.

Sharma, Y. C., B. Singh, and J. Korstad. 2009. "High Yield and Conversion of Biodiesel from a Nonedible Feedstock (*Pongamia Pinnata*)." *Journal of Agricultural and Food Chemistry* 58 (1):242–247.

Sharma, Y. C., B. Singh, and J. Korstad. 2010. "Application of an Efficient Nonconventional Heterogeneous Catalyst for Biodiesel Synthesis from Pongamia Pinnata Oil." *Energy and Fuels* 24 (5):3223–3231.

Shaw, J.-F., S.-W. Chang, S.-C. Lin, T.-T. Wu, H. Y. Ju, and C. C. Akoh. 2008. "Continuous Enzymatic Synthesis of Biodiesel with Novozym 435." *Energy & Fuels* 22 (2):840–844.

Sheehan, J., T. Dunahay, R. Benemann, G. Roessler, and C. Weissman. 1998. "A Look Back at the U.S. Department of Energy's Aquatic Species Program: Biodiesel from Algae; Close-Out Report." National Renewable Energy Laboratory.

Shi, W., J. Li, B. He, F. Yan, Z. Cui, K. Wu, L. Lin, X. Qian, and Y. Cheng. 2013. Biodiesel Production from Waste Chicken Fat with Low Free Fatty Acids by an Integrated Catalytic Process of Composite Membrane and Sodium Methoxide. *Biosource Technology* 139:316–322.

Shimada, Y., Y. Watanabe, T. Samukawa, A. Sugihara, H. Noda, H. Fukuda, and Y. Tominaga. 1999. "Conversion of Vegetable Oil to Biodiesel Using Immobilized *Candida Antarctica* Lipase." *Journal of the American Oil Chemists' Society* 76 (7):789–793.

Shin, H.-Y., S.-H. Lee, J.-H. Ryu, and S.-Y. Bae. 2012. "Biodiesel Production from Waste Lard Using Supercritical Methanol." *Journal of Supercritical Fluids* 61:134–138.

Silva, C. C. C. M., N. F. P. Ribeiro, M. M. V. M. Souza, and D. A. G. Aranda. 2010. "Biodiesel Production from Soybean Oil and Methanol Using Hydrotalcites as Catalyst." *Fuel Processing Technology* 91 (2):205–210.

Singh, Y., and H. D. Kumar. 1992. "Lipid and Hydrocarbon Production by Botryococcus Spp. Under Nitrogen Limitation and Anaerobiosis." *World Journal of Microbiology and Biotechnology* 8 (2):121–124.

Sinha, S., A. K. Agarwal, and S. Garg. 2008. "Biodiesel Development from Rice Bran Oil: Transesterification Process Optimization and Fuel Characterization." *Energy Conversion and Management* 49 (5):1248–1257

Sivasamy, A., K. Y. Cheah, P. Fornasiero, F. Kemausuor, S. Zinoviev, and S. Miertus. 2009. "Catalytic Applications in the Production of Biodiesel from Vegetable Oils." *ChemSusChem* 2:278–300.

Soumanou, M. M., and U. T. Bornscheuer. 2003. "Improvement in Lipase-Catalyzed Synthesis of Fatty Acid Methyl Esters from Sunflower Oil." *Enzyme and Microbial Technology* 33 (1):97–103.

Spolaore, P., C. Joannis-Cassan, E. Duran, and A. Isambert. 2006. "Commercial Applications of Microalgae." *Journal of Bioscience and Bioengineering* 101 (2):87–96.

Szczesna Antczak, M., A. Kubiak, T. Antczak, and S. Bielecki. 2009. "Enzymatic Biodiesel Synthesis—Key Factors Affecting Efficiency of the Process." *Renewable Energy* 34 (5):1185–1194.

Taher, H. E. 2009. "Enzymatic Production of Biodiesel from Fats Extracted from Lamb Meat Using Supercritical CO_2." United Arab Emirates University, College of Engineering.

Taher, H. E. 2014. "Development of a Process for Producing Biodiesel from Microalgae Lipids with Supercritical Carbon Dioxide Extraction and Enzyme-Catalyzed Transesterification." Dissertation/PhD, United Arab Emirates University, College of Engineering.

Taher, H., S. Al-Zuhair, A. Al-Marzouqi, and I. Hashim. 2011. "Extracted Fat from Lamb Meat by Supercritical CO_2 as Feedstock for Biodiesel Production." *Biochemical Engineering Journal* 55 (1):23–31.

Taher, H., S. Al-Zuhair, A. H. Al-Marzouqi, Y. Haik, and M. Farid. 2014a. "Effective Extraction of Microalgae Lipids from Wet Biomass for Biodiesel Production." *Biomass and Bioenergy* 66:159–167.

Taher, H., S. Al-Zuhair, A. H. Al-Marzouqi, Y. Haik, and M. Farid. 2014b. "Enzymatic Biodiesel Production of Microalgae Lipids under Supercritical Carbon Dioxide: Process Optimization and Integration." *Biochemical Engineering Journal* 90:103–113.

Tan, C.-S., S.-K. Liang, and D.-C. Liou. 1988. "Fluid-Solid Mass Transfer in a Supercritical Fluid Extractor." *Chemical Engineering Journal* 38 (1):17–22.

Tan, K. T., M. M. Gui, K. T. Lee, and A. R. Mohamed. 2010. "An Optimized Study of Methanol and Ethanol in Supercritical Alcohol Technology for Biodiesel Production." *Journal of Supercritical Fluids* 53 (1–3):82–87.

Taufiqurrahmi, N., A. R. Mohamed, and S. Bhatia. 2011. "Production of Biofuel from Waste Cooking Palm Oil Using Nanocrystalline Zeolite as Catalyst: Process Optimization Studies." *Bioresource Technology* 102 (22):10686–10694.

Varma, M. N., and G. Madras. 2007. "Synthesis of Biodiesel from Castor Oil and Linseed Oil in Supercritical Fluids." *Industrial & Engineering Chemistry Research* 46 (1):1–6.

Varma, M. N., P. A. Deshpande, and G. Madras. 2010. "Synthesis of Biodiesel in Supercritical Alcohols and Supercritical Carbon Dioxide." *Fuel* 89 (7):1641–1646.

Verleyen, T., U. Sosinska, S. Ioannidou, R. Verhe, K. Dewettinck, A. Huyghebaert, and W. De Greyt. 2002. "Influence of the Vegetable Oil Refining Process on Free and Esterified Sterols." *Journal of the American Oil Chemists' Society* 79 (10):947–953.

Vicente, G., M. Martínez, and J. Aracil. 2004. "Integrated Biodiesel Production: A Comparison of Different Homogeneous Catalysts Systems." *Bioresource Technology* 92 (3):297–305.

Vieitez, I., C. da Silva, G. R. Borges, F. C. Corazza, J. V. Oliveira, M. A. Grompone, and I. Jachmanián. 2008. "Continuous Production of Soybean Biodiesel in Supercritical Ethanol–Water Mixtures." *Energy & Fuels* 22 (4):2805–2809.

Vijayaraghavan, K., and K. Hemanathan. 2009. "Biodiesel Production from Freshwater Algae." *Energy & Fuels* 23 (11):5448–5453.

Voltolina, D., M. del Pilar Sanchez-Saavedra, and L. Maria Torres-Rodriguez. 2008. "Outdoor Mass Microalgae Production in Bahia Kino, Sonora, NW Mexico." *Aquacultural Engineering* 38 (2):93–96.

Vyas, A. P., J. L. Verma, and N. Subrahmanyam. 2010. "A Review on Fame Production Processes." *Fuel* 89 (1):1–9.

Wang, W. G., D. W. Lyons, N. N. Clark, M. Gautam, and P. M. Norton. 2000. "Emissions from Nine Heavy Trucks Fueled by Diesel and Biodiesel Blend without Engine Modification." *Environmental Science & Technology* 34 (6):933–939.

Wang, L., W. Du, D. Liu, L. Li, and N. Dai. 2006. "Lipase-Catalyzed Biodiesel Production from Soybean Oil Deodorizer Distillate with Absorbent Present in tert-Butanol System." *Journal of Molecular Catalysis B: Enzymatic* 43 (1–4):29–32.

Xie, W., H. Peng, and L. Chen. 2006. "Calcined Mg-Al Hydrotalcites as Solid Base Catalysts for Methanolysis of Soybean Oil." *Journal of Molecular Catalysis A: Chemical* 246 (1–2):24–32.

Xu, Y., W. Du, and D. Liu, and J. Zeng. 2003. "A Novel Enzymatic Route for Biodiesel Production from Renewable Oils in a Solvent-Free Medium." *Biotechnology Letters* 25 (15):1239–1241.

Xu, W. D., J. Zeng, and Y. D. Liu. 2004. "Conversion of Soybean Oil to Biodiesel Fuel Using Lipozyme TL IM in a Solvent-Free Medium." *Biocatalysis and Biotransformation* 22 (1):45–48.

Xu, Y., W. Du, and D. Liu. 2005. "Study on the Kinetics of Enzymatic Interesterification of Triglycerides for Biodiesel Production with Methyl Acetate as the Acyl Acceptor." *Journal of Molecular Catalysis B: Enzymatic* 32 (5–6):241–245.

Yin, C.-Y. 2011. "Prediction of Higher Heating Values of Biomass from Proximate and Ultimate Analyses." *Fuel* 90 (3):1128–1132.

Yori, J. C., S. A. D'Ippolito, C. L. Pieck, and C. R. Vera. 2007. "Deglycerolization of Biodiesel Streams by Adsorption over Silica Beds." *Energy & Fuels* 21 (1):347–353.

Zabeti, M., W. M. A. W. Daud, and M. K. Aroua. 2009. "Activity of Solid Catalysts for Biodiesel Production: A Review." *Fuel Processing Technology* 90 (6):770–777.

Zaks, A., and A. M. Klibanov. 1984. "Enzymatic Catalysis in Organic Media at 100 Degrees C." *Science* 224:1249–1251.

Zhang, Y., M. A. Dube, D. D. McLean, and M. Kates. 2003. "Biodiesel Production from Waste Cooking Oil: 1. Process Design and Technological Assessment." *Bioresource Technology* 89 (1):1–16.

Zhang, X., J. Li, Y. Chen, J. Wang, L. Feng, X. Wang, and F. Cao. 2009. "Heteropolyacid Nanoreactor with Double Acid Sites as a Highly Efficient and Reusable Catalyst for the Transesterification of Waste Cooking Oil." *Energy & Fuels* 23 (9):4640–4646.

Zheng, S., M. Kates, M. A. Dube, and D. D. McLean. 2006. "Acid-Catalyzed Production of Biodiesel from Waste Frying Oil." *Biomass and Bioenergy* 30 (3):267–272.

Zheng, Y., J. Quan, X. Ning, L.-M. Zhu, B. Jiang, and Z.-Y. He. 2009. "Lipase-Catalyzed Transesterification of Soybean Oil for Biodiesel Production in tert-Amyl Alcohol." *World Journal of Microbiology and Biotechnology* 25 (1):41–46.

Zhou, W., S. K. Konar, and D. G. B. Boocock. 2003. "Ethyl Esters from the Single-Phase Base-Catalyzed Ethanolysis of Vegetable Oils." *Journal of the American Oil Chemists' Society* 80 (4):367–371.

Zullaikah, S., C.-C. Lai, S. R. Vali, and Y.-H. Ju. 2005. "A Two-Step Acid-Catalyzed Process for the Production of Biodiesel from Rice Bran Oil." *Bioresource Technology* 96 (17):1889–1896.

Index